# CARBON CHANGE

# CARBON CHANGE

## CANADA ON THE BRINK OF DECARBONIZATION

DENNIS McCONAGHY

Publisher: Kwame Scott Fraser | Acquiring editor: Kathryn Lane | Editor: Dominic Farrell
Cover designer: Karen Alexiou
Cover image: Jens Johnsson on Unsplash

**Library and Archives Canada Cataloguing in Publication**

Title: Carbon change : Canada on the brink of decarbonization / Dennis McConaghy.
Names: McConaghy, Dennis, 1952- author.
Description: Includes index.
Identifiers: Canadiana (print) 20220254974 | Canadiana (ebook) 20220255113 | ISBN 9781459750517 (softcover) | ISBN 9781459750524 (PDF) | ISBN 9781459750531 (EPUB)
Subjects: LCSH: Energy transition—Canada. | LCSH: Energy transition—Canada—Costs. | LCSH: Energy policy—Canada. | LCSH: Carbon dioxide mitigation—Canada. | LCSH: Carbon dioxide mitigation—Canada—Costs. | LCSH: Climate change mitigation—Canada.
Classification: LCC HD9502.C32 M2937 2022 | DDC 333.790971—dc23

We acknowledge the support of the Canada Council for the Arts and the Ontario Arts Council for our publishing program. We also acknowledge the financial support of the Government of Ontario, through the Ontario Book Publishing Tax Credit and Ontario Creates, and the Government of Canada.



Printed and bound in Canada.

Dundurn Press
1382 Queen Street East
Toronto, Ontario, Canada M4L 1C9
dundurn.com, @dundurnpress

*To children everywhere, in the hope that the possibility of living in a world that can evolve more rationally and safely still exists.*

# Contents

# Introduction

After publishing my first two books, *Dysfunction* (2017), about the drawn-out demise of the Keystone XL pipeline, and its sequel, *Breakdown* (2019), about the links between resource development, climate change, Indigenous rights, and Alberta alienation, I did not expect to write a third; it seemed to me that I had substantially contributed all I had to offer. However, since 2019 I have watched the world's approach to dealing with the risk of climate change moving in a direction that I believe needs reconsidering.

The United Nations process designed to reduce the concentration of greenhouse gases (GHGs) in the atmosphere, established in the early 1990s, has been adopted by virtually every nation. It relies on the findings of an extensive group of scientists and other scholars to assess the physical and economic dimensions of climate change. That process, the only one of its scope, has adopted as orthodoxy that there is an urgent need to radically decarbonize the developed world's energy systems, with the aim of stabilizing global average temperatures at a level no more than 1.5°C above pre-industrial norms. To be clear, to decarbonize means that hydrocarbons could no longer be produced or consumed — full stop. The related concept of "net-zero emissions" leaves open the option of hydrocarbon production and consumption, provided that any related GHG emissions are not released into the atmosphere. This possibility might be currently affordable — the claim is debatable — in a very few contexts, such as would be found in coal-based electric generation facilities. Beyond that, the current status of the technology is highly challenged economically.

Some proponents of this transformation contend that even with decarbonization, energy will remain as affordable and available as it is today, a situation that will allow our economies to remain virtually unchanged. We could enjoy continued growth and the same living standards, but we would not be adding additional GHGs to the atmosphere. Others who advocate decarbonization make no such claim but contend that no price, even a significant negative impact on human welfare, is too great to reduce emissions.

Climate change risk is real and serious, as is evidenced by its impacts on the world's oceans, on its forests and farmland, and on biodiversity in every ecosystem, and it is clearly attributable, for the most part, to human activity. There has been a measurable rise in global temperatures since pre-industrial times; by the end of 2021, that increase was on the order of 1°C. The United Nations' Intergovernmental Panel on Climate Change (IPCC) Sixth Assessment Report from its Working Group 1 states, "The frequency and intensity of hot extremes have increased and those of cold extremes have decreased on the global scale from 1950." It concludes that it is an "established fact" that human-caused emissions of greenhouse gases "have led to an increased frequency and/or intensity of some weather and climate extremes."

My passionate belief is that climate change must be addressed globally, but with policies and actions consistent with both our understanding of the physical risk and the costs associated with attempting to lessen that risk and adapting to whatever changes occur as a result of climate change. The world should reconsider decarbonization and focus on an alternative method of dealing with the risk of climate change, one based on adapting to a level of global temperature increase higher than 1.5°C, and more likely on the order of 3°C.

Of course, for this to occur, the UN process for dealing with climate change would have to be reinvented. Currently, that process allocates emissions reductions to those developed countries that are major emitters, while allowing others, such as China and India, to be "free riders," not bearing a proportionate share of the costs of mitigating climate change risk. This is not working. Instead, the process should be predicated on carbon pricing set via carbon taxes on emitted GHGs and applied consistently across the world's developed economies. Not only would this approach avoid the free rider problem, which is completely ignored in the current approach, but the revenue from carbon taxes could fund adaptation in those countries least culpable for the accumulation of emissions in the atmosphere, historically and currently, and most vulnerable to the impacts of climate change. Carbon pricing would be constrained by the social cost of carbon, that is, the net cost arising from climate change, accounting for the benefits of continued hydrocarbon consumption and higher GHG concentrations in the atmosphere. A cost/benefit analysis approach would create a more economically optimal outcome for managing the climate change risk than unconditional adherence to decarbonization.

○ ○ ○

Since my last book was published in 2019, the world has been profoundly impacted by the advent of the Covid-19 pandemic. The world's reaction to it offers, according to many, an example of how we ought to react to the risk of climate change, most notably strict compliance with government interventions, regardless of the economic consequences. Climate change and the Covid-19 pandemic have in common the fact that they are global in scope. However, I take issue with the proposition that Covid-19 and climate change both pose immediate existential risks to humanity, if by that we mean a situation in which humanity faces imminent extinction, and therefore ought to be addressed with similarly extreme measures.

Governments have dealt with the pandemic by radically intervening in how their citizens are allowed to lead their lives (lockdowns, quarantines, masks, vaccination mandates), and populations have mostly complied with the regulations imposed and requests made by governments, all regardless of the sacrifices required, whether in terms of economic contraction or personal liberty. This was done to minimize deaths and preserve hospital capacity. It has remained a matter of contentious debate whether that represented the best response prior to the advent of vaccines. Not even in the earliest months of dealing with Covid-19 was it ever contended that the ultimate mortality rates would exceed much more than 2 percent. And that limit has persisted at least up to the end of 2021. The Covid-19 experience to date has not been an existential crisis.

The term *existential risk* is not used in any of the actual IPCC reports comprising its Sixth Assessment Report of Working Group 1, notwithstanding the panel's resort to the florid language of "code red for humanity" in its *Summary for Policy Makers*. The capacity to deal with risk is reaffirmed by the IPCC's actual technical reports. To date, the synthesis of physical and economic consideration will not be available until late 2022, when the IPCC completes the full Sixth Assessment Report. Resort to language that creates panic and overreaction, whether regarding Covid-19 or climate change, is not constructive.

The goal in dealing with climate change is to set policy objectives consistent with the risk and economic capacity to afford both the requisite reduction of that risk and adaptation to whatever changes do occur, and then take the necessary actions to meet those objectives. What was justifiable for Covid-19 is unlikely to be appropriate for dealing with climate change, regardless of how claims of "existential crisis" are ascribed to the climate change risk.

From 2020 through to the end of that year, the world grappled with great uncertainty. In March 2020 no one could claim definitively when, or how, the pandemic could be resolved. How long would lockdowns persist? Could the world economically sustain the lockdown indefinitely? What level of herd immunity might populations achieve? Would we eventually have to resort to ending lockdowns and letting mass infection sort out the rest? How many humans would become infected, and how many would die? Of course, the world was largely spared coming to terms with the most difficult, almost Darwinian, trade-offs, implicit in these questions, thanks to the advent of vaccines — a technological breakthrough that proved largely affordable for developed countries to deploy. By mid-2022, the world was making an uneasy but inexorable return to normalcy.

For the climate change risk, we have no affordable technological breakthrough, to date, that could parallel the effect of vaccines on the Covid-19 pandemic. Climate change also contrasts with the pandemic in that it is a true "tragedy of the commons" — the situation where a shared resource creates an incentive for some who consume or utilize it to let others who are also reliant on it to bear the cost of maintaining that resource — such that collective, coordinated, and enforced action is the only way forward to deal with its intrinsic perverse nature. The UN process has utterly failed to achieve that type of action after almost forty years. There is no equivalent to quarantines, let alone vaccines, as yet, when it comes to the lessening of climate change risk. The basic question persists: Is decarbonization the best course for trying to do so?

In early 2022, another cataclysm occurred: the Russian invasion of Ukraine. And with it, energy security was restored as a pre-eminent consideration for how both energy and climate policy would unfold across the developed economies of the West. How much longer could dependence on Russian hydrocarbons, most significantly natural gas, be countenanced, especially for Central Europe? And how much would North America, both Canada and the United States, step up its hydrocarbon production to physically displace existing Russian supply, even if the inescapable consequence was that carbon emissions would grow materially over medium and longer term? How much should energy security be valued, whether in terms of reduced leverage accorded Russia over Central Europe or higher economic integration of North American and Europe in energy supply? Over the short and medium terms, for Europe, the issue is not whether hydrocarbons are required (they inescapably are), but rather if it can continue to be dependent on Russian supply in the context of their unjustified disruption of the world order and fundamental barbarism. These questions are now front and centre in the spring of 2022, creating a brutal realism to the fundamental arguments of optimal climate policy, let alone the merits of decarbonization.

o o o

The book has two parts. It begins with a review of the current argument for decarbonization, the consequences of trying to achieve it in terms of our fundamental energy systems, and a review of the UN process that has been driving this course forward. I clarify crucial concepts, from carbon budgets to climate sensitivity to the treatment of tail risk — events that occur at either end of a normal distribution curve and have low probability but potentially high impact — related to climate change. Then, the specific developments in Canada and the United States over the period from 2019 to late 2021 will be discussed to show how climate policy has impacted energy development in both countries.

In part two, I offer specific recommendations for reconsidering decarbonization. In the latter section of chapter 8, with elaboration in chapters 9 and 10, I offer specific actions to bring that reconsideration into effect. I emphasize how Canada should relate to this fundamental policy question, as a country that has tried consistently for thirty years to relate to the UN process, regardless of the cost imposed on itself, especially relative to other developed economies that are Canada's major trading partners. I appraise whether reconsidering decarbonization is even possible at this time, when most developed economies are committed to it, along with powerful financial and media elites. It is my view that such a reconsideration can, in fact, occur, so long as free markets and functioning democracies still exist. If that continues to hold true, the costs of energy transition to be borne by the electorates of developed economies will still require validation by the democratic process. Such validation will be based, I hope, on transparent and credible analysis of the net costs of all alternatives to deal with the climate change risk. Furthermore, the net cost any particular country bears must be fair, relative to what others bear. This is especially important for Canada, which faces not only higher costs to achieve incremental — that is, increased — emissions reductions and will be challenged to sacrifice its hydrocarbon industry with no assurance that other hydrocarbon producers would do the same.

But now based on the events of early 2022, any commitment to decarbonization by North America particularly must also account for the energy security implications in a more urgent and serious manner than would have been the case previously, literally over the entire course of the current UN process.

This book does not embrace climate change denialism, but it is a passionate advocacy of a reconsideration of optimal policies to reduce the expected risk of climate change.

# PART ONE

# The Climate and Energy Context

# CHAPTER 1

# The UN Climate Process: How We Got to Decarbonization

The UN formed its Intergovernmental Panel on Climate Change (IPCC) in 1988 to prepare "a comprehensive review and recommendations with respect to the state of knowledge of the science of climate change; the social and economic impact of climate change; and potential response strategies and elements for inclusion in a possible future international convention on climate." This occurred shortly after Dr. James Hansen, director of NASA's Institute for Space Studies, made his seminal testimony before the U.S. Senate Energy and Natural Resources Committee, proclaiming the onset of "global warming."

Participation in the IPCC is open to all member countries of the World Meteorological Organization (WMO) and the UN, and today, virtually every country in the world has representation on the IPCC. Its main task is to complete various assessment reports, which have evolved to include major components from three "working groups," examining, respectively, the physical science of climate change, the effects of climate change on nature and society, and methods for adaptation and mitigation. Each group issues its own report, prepared by lead authors who are nominated by governments and selected on the basis of evidence of active participation in a relevant research area. Each group is accountable for the content of specific chapters, relying primarily on review of relevant peer-reviewed literature. Subsequent review occurs with a broader array of experts for most subject areas, with the ultimate objective of providing a balanced synthesis of the present state of scientific knowledge and related uncertainties relating to global climate change.

Significantly, the actual IPCC panel is not comprised of content experts, but of individuals nominated by their respective governments at their discretion, presumably to reflect their countries' climate priorities as much as to ensure the intellectual integrity of the entire process. The panel ultimately creates a "Summary for Policy Makers" (SPM). In many respects, this summary serves as the key deliverable of each assessment report cycle, and gains wide media and political attention, serving to frame the deliberations of the UN process itself. The SPM is as much a politically compromised document as a dispassionate summation of scientific findings.

o o o

In 1992, at the "Earth Summit" in Rio de Janeiro convened by the United Nations, the UN Framework Convention on Climate Change (UNFCCC) was established to reduce emissions and foster adaptation to change. Initially, 154 countries signed on, and as of 2021 there were 197 parties to the convention, including Canada and the United States. A key component of the convention was the commitment to an annual meeting, which has come to be known as the Conference of the Parties (COP), to advance both policy and action to meet the mandate. But more substantively, UNFCCC enshrined the principle of "common but differentiated responsibilities and respective capabilities." In other words, it acknowledged both that the developed world was culpable for climate change and that the developing world was constrained in its capacity to help diminish the effects. The process internalized a North-South dynamic, notwithstanding its stated purpose to "stabilize greenhouse gas concentrations at a level that would prevent dangerous anthropogenic interference with the climate system ... within a time-frame sufficient to allow ecosystems to adapt naturally to climate change, to ensure that food production is not threatened, and to enable economic development to proceed in a sustainable manner." Developed countries expected to carry the burden of dealing with the climate change risk were itemized in Annex I on the convention's original treaty document. And perhaps as profoundly, the UNFCCC decided to accept a consensus-based decision-making process, not one prioritizing those who would carry the burden of mitigation.

The UNFCCC's ambitions culminated in 1997 with the negotiation of the Kyoto Protocol, which required developed economies to reduce emissions, by 2012, to 5 percent below 1990 levels. This reduction level was viewed at the time as consistent with the 2°C temperature containment objective, although that was not explicitly cited. The protocol created an international trading system whereby countries with surplus emission allowances could "sell" them to other countries — one of the North-South adjustments that typified the ethos of the protocol. Kyoto did not expect, let alone compel, major current emitters, such as China or India, to reduce their emissions. Although under the Clinton administration the United States had signed the protocol, Congress never ratified it, mostly because the agreement required no reduction commitment from China. Other countries, including Canada, ratified the protocol but later withdrew, essentially when they recognized that the required emissions reductions were never going to materialize within the required timeline of the first reporting period in 2012. From the protocol's inception until the end of its anticipated first commitment period in 2012, global emissions grew steadily.

The 2009 COP took place in Copenhagen, burdened by the enormously ambitious objective of replacing Kyoto with a new binding global agreement to realize the stated goal of the UNFCCC itself, to reduce emissions and adapt to the effects of climate change. The parties achieved no such agreement. They could not overcome fundamental disagreements about how the reductions ought to be allocated, the aggregate number of those reductions, or financial assistance for developing countries to cope with the impact of climate change. The parties settled on a non-binding document that was "taken note of," not adopted — UN speak for something more than total meltdown.

The Copenhagen Accord acknowledges that global temperatures should not increase by 2°C (3.6°F) above pre-industrial levels. Representatives from developing countries sought a target of 1.5°C (2.7°F). Still, certain countries did embrace the 2°C global temperature containment goal explicitly for the first time, Canada and the United States among them. Both Canada and the United States, also, in keeping with that goal, accepted an emissions reduction objective of 5 percent below 2005 emission levels by 2030, albeit on a non-binding basis. But thereafter, not surprisingly, Canada's emissions proceeded to grow steadily up to November 2015. The targets never reconciled with reality because of energy markets at odds with climate aspirations, for both Canada and the United States.

COP21, held in Paris in November 2015, was to represent the culmination of the Obama administration's climate diplomacy. In the Kyoto Protocol and during subsequent efforts at Copenhagen, the parties had allocated emissions reductions to each developed country in order to reach sufficient global reductions consistent with a specific temperature containment goal. For the Paris Accord — an agreement, not a treaty — countries offered pledges of their own devising, specifying their emissions reduction goals out to 2030. Whatever the sum of those pledges, they would be accepted, even if the sum could not achieve 2°C containment — which was presumably the explicit objective of the Paris Accord itself. Canada and the United States maintained their Copenhagen reduction goals; however, China and India's positions remained problematic. Neither country submitted any explicit emissions goal, only statements that their countries would reach "peak" emissions by 2030. This clearly underwhelming outcome was the best the UNFCCC could practically achieve without the Paris talks breaking down entirely. According to apologists of the UNFCCC and Obama administration, Paris was still a laudable foundation to build on. Implicitly, they expected the parties' emissions reductions to grow more ambitious over time.

Perhaps most importantly, the Paris Accord foreshadowed the emergence of a 1.5°C temperature containment target — essentially, the call to decarbonize. The accord defines its basic objective as "holding the increase in the global average temperature well below 2°C above pre-industrial levels, and to pursue efforts to limit the temperature increase to 1.5°C" (Article 2.1). The document also introduces the concept of "zero net emissions." In its AR5 Assessment Report, the IPCC had projected a global temperature increase of 3 or 4°C, unless the world embraced new and more drastic emissions reductions. At best, the actual Paris emissions reduction targets, if achieved, promised temperature containment on the order of 2.7°C. Carbon pricing remained peripheral to the Paris Accord. The parties tried to perfect a system for a global market for the "buy/sell" of offsets, but those efforts were incomplete, to be carried forward to subsequent years. From the Copenhagen climate meeting, the accord retained the concept of transferring money from North to South — specifically, US$100 billion annually, to deal with climate change impacts, by 2020. (As of November 2021, the developed world had clearly not met that commitment. Even the measurement of what transfers or financings had been achieved since Paris would prove to be contentious.) Finally, the Paris Accord was still fundamentally non-binding, with no explicit enforcement mechanism or financial consequence for non-compliance.

As of November 2017, sufficient countries had ratified the Paris Accord for it to come into effect in November 2020. However, in June 2017, U.S. president Donald Trump announced his intention to withdraw the United States. That meant roughly 11 percent of global GHG emissions were no longer subject to the terms of the accord. Trump claimed that the accord had been unfair to the United States, demanding excessive emissions reductions that would harm the nation's economy, while other major world economies, most notably China, enjoyed exemptions from any comparable obligations. Again, China had pledged merely that its emissions would peak by 2030, an imbalance justified by its comparatively small historic culpability for atmospheric accumulations of GHGs and its status as a developing nation. Because the Obama administration had treated the Paris Accord as an agreement, not a formal treaty requiring Senate ratification, the accord had always been vulnerable to some future administration withdrawing, and Trump did just that. Meanwhile, Obama's original attempt to finesse the accord around the U.S. Congress had demonstrated the country's tenuous political support for any international reduction commitment, at least in the Congress and Senate. Still, ironically, with no federal climate legislation, and without participating in the Kyoto Accord or Paris Accord, the United States saw its GHG emissions decline to 5,500 gigatonnes per year by the end of 2019 — in other words, by 13 percent since 2005 — primarily thanks to substituting natural gas for coal in electric generation via market forces, not government decisions.

In 2018, at the request of the UNFCCC, the IPCC issued a special report on 1.5°C temperature containment. One of its primary findings was that, in order to achieve that goal, the world would have to reach net-zero emissions by around 2050. Otherwise, the report warned, temperature increase would exceed 1.5°C containment around 2030 and would exceed 3°C by midcentury. If that occurred, the IPCC expressed "high confidence" that tropical areas would face frequent severe heat waves and that the world would suffer from other extreme weather events and mounting impacts on global biodiversity. The report emphasized that these impacts would fall disproportionately on the poor and vulnerable. According to the report, the difference between 1.5°C containment and 2°C containment brought specific risks, including ten times likelier Arctic ice-free summers, a 23 percent increase in the world population exposed to "highly unusual" hot days, one hundred million more people exposed to severe drought, double the percentage of plants and animals losing their range of habitat, the disappearance of most coral reefs, and a 15 percent increase in global population exposed to sea-level rise — up to eighty million people. The report's most alarmist contention was that, on the other side of 1.5°C, the world could reach a "tipping point," after which sudden and calamitous climate change would rapidly speed up. As for cost, the IPCC's special report estimated that decarbonizing would require $3 trillion to $3.5 trillion per year over the next three decades. The report included no cost/benefit analysis on the relative merits of decarbonizing versus adapting to other temperature containment objectives.

Even with a Trump administration indifferent to the UN process, this galvanized more developed countries to embrace the net-zero emissions goals, as least aspirationally. At the next COP meeting, in Glasgow in 2021, that goal would presumably manifest in more ambitious national targets than might otherwise have been expected. How far and how credibly would the developed world go before the actual deliberations of COP26 in Glasgow in November 2021?

Since the inception of the UNFCCC, it has not relied on conventional cost/benefit analysis, integrating economic and physical considerations, to provide guidance for policy makers. The UNFCCC process emphasizes potential catastrophic impacts and existing regional and intergenerational inequities that might otherwise be undervalued by accepted cost/benefit methodology, most notably via discounting and the conversion of impacts to financial equivalents; these factors are more ignored than explicitly rejected as a basis for decision making. Conceptually, cost/benefit analysis amounts to a netting out of the costs of climate change mostly represented as damage, versus the costs of abating those damages. What is the net cost or benefit? How is the net benefit optimized? At what level of temperature containment?

# CHAPTER 2

# The Orthodoxy: From Climate Sensitivity to Carbon Budgets to Decarbonization

In 2021 the political leadership of most of the developed world had officially reached consensus on how to deal with climate change risk: the world's energy systems must be decarbonized. This, they agreed, was the only path available if the world wanted to contain the rise of global average temperature to no more than 1.5°C above the pre-industrial average. Indeed, the word *consensus* doesn't quite convey how this — still eminently debatable — response to dealing with climate change risk has become the politically correct orthodoxy, such that any skepticism is deemed by most political elites in the developed world as either immoral, politically blighted, or just plain stupid. Yet, I am skeptical. While I believe that climate change requires a proportionate and credible global response, I do not consider global decarbonization necessarily the optimal policy objective; moreover, it is an especially reckless objective for Canada and the United States to embrace — Canada because its economy relies significantly on hydrocarbon production, and the United States due to its geopolitical and economic position in the world.

Let me reiterate: the developed world urgently needs to reconsider its commitment to decarbonizing. Here I must note the distinction between a decarbonized world and one with "net-zero emissions," a term that is often invoked by those who wish to make the enormity and extreme implications of decarbonization appear more palatable and feasible. As mentioned earlier, decarbonization demands that no hydrocarbons are combusted, period, while in a world with net-zero emissions, humans can still emit GHGs, as long as we sequester them before they enter the atmosphere or remove them from the atmosphere once they've been emitted into it. Advocates for net-zero emissions champion the potential of carbon capture or storage (CCS) — the term given to the processes and technologies that sequester emissions, usually in natural reservoirs underground, and thereby keep them out of the atmosphere. As already noted, it is only in a few select circumstances that today's technology can accomplish CCS at a scale sufficiently large enough to accommodate large-scale emissions, and at affordable cost.

Carbon offsets, despite the claims made for them by their supporters, cannot facilitate net-zero emissions, since they merely move permissions to emit between countries. Typically, a developed country, subject to emissions reduction obligations according to conventions such as the current Paris Accord, can buy permission to emit GHGs from less-developed countries not subject to the same obligations. For instance, North American coal-based electric power generators could fund measures to reduce methane emissions from natural gas pipelines in Asia or Africa rather than reducing their own emissions. It is clear why such measures will never help accomplish net-zero emissions and a decarbonized world: they do not alter the aggregate volume of emissions globally but, at best, merely reduce the cost of achieving a specific level of emissions reduction.

Since technologies that might sequester or remove GHGs are not currently viable, economically or at sufficient scale, the UN process demands, de facto, that humanity cease combusting hydrocarbons. This extreme position is what I mean, then, by "decarbonization."

o o o

In August 2021, the UN's Intergovernmental Panel on Climate Change Working Group 1 released its Sixth Assessment Report (AR6) on the state of climate change. AR6 comprises mostly technical reports; it is a collation of the most recent climate change data available and was designed to be used to directly inform the mandate to decarbonize. The report cites "unequivocal evidence" that the world has warmed more since the 1950s than it did in the previous hundreds or thousands of years, and "at rates unprecedented in at least the last two thousand years." The evidence of human culpability for this change has become more conclusive in each IPCC assessment report, and the authors of AR6 state that that evidence "is even stronger in this assessment, including for regional scales and for extremes." Indeed, the report calls it an "established fact" that human-caused emissions of GHGs "have led to an increased frequency and/or intensity of some weather and climate extremes."

The temperature rise will inevitably continue, the report says, but can almost certainly be "limited through rapid and substantial reductions in global GHG emissions." Without adequate emissions reductions, on the other hand, global temperature increase will exceed 1.5°C in the early 2030s and 2°C before the end of the century. Climate risks vary across the world, but all regions are projected to experience changes even with 1.5°C warming, and incrementally more pronounced ones with increased warming. After 2°C warming, the projections warn, the fire season across North America "expands dramatically." At 4°C, heavy precipitation events previously seen once in ten or more years will, certainly, "become more frequent and more intense than in the recent past, on the global scale ... and in all continents and AR6 regions." The report predicts another century of "near-linear dependence of global temperature on cumulative GHG emissions," meaning there is no evidence of an approaching tipping point at which global temperature races suddenly and irreversibly upward.

Secretary-General of the United Nations António Guterres offered a predictable, politically correct response to the AR6, one that ignored the economic implications of the technical reports' actual language. He stated:

> Today's IPCC Working Group 1 Report is **a code red for humanity**. The alarm bells are deafening, and the evidence

> is irrefutable: greenhouse gas emissions from fossil fuel burning and deforestation are choking our planet and putting billions of people at immediate risk. Global heating is affecting every region on Earth, with many of the changes becoming irreversible.... We are at imminent risk of hitting 1.5 degrees in the near term. The only way to prevent exceeding this threshold is by urgently stepping up our efforts and pursuing the most ambitious path. (Emphasis mine)

Guterres urged the world to "act decisively now to keep 1.5 alive." The language in his response was dramatic, though not as extremist as that found in an article published in the *Guardian*: "As a verdict on the climate crimes of humanity, the new Intergovernmental Panel on Climate Change report could not be clearer: guilty as hell."

These formulations lead logically to decarbonization: if we accept that emitting GHGs into the atmosphere will, beyond certain levels, cause a higher risk of catastrophic consequences, including — potentially — some risk of threatening human existence itself, then, given the accumulation of such emissions already in the atmosphere, complete decarbonization is the only way to mitigate, if not avert, that risk. Make no mistake: containing global temperature rise to 1.5°C would require decarbonization as soon as practically possible, if existing climate models have any credibility.

This argument ignores the fact that adapting to some higher level of global temperature increase may offer a higher net benefit to the global economy, even taking into account the actual (very low) probability of catastrophes so extreme as to justify the term *existential risk*. All political leaders, domestically and globally, should be asking what level of temperature containment offers the highest net benefit to their citizens. But doing so is currently not an option, because the proponents of decarbonization insist that containing 1.5°C is the only viable option for humanity, using rhetoric that frames the elimination of hydrocarbons from the global economy as a moral imperative.

It is this absolutist position with which I take issue. The official and loudest voices in the climate-policy world have limited the conversation by moralizing, and by considering only the potential damages that climate change might cause. This is an inadequate substitute for dispassionate cost/benefit analysis. Assessing climate change risk, like assessment of any risk, requires examining and comparing probability and consequences. That should not be forgotten, especially in the context of decarbonization. Other climate policies and tactics may better diminish climate change risk, while avoiding undue destruction of the world economy. And, despite the developed world's ostensible commitment to decarbonization, as of the end of 2021 its democratic components had shown themselves unable to unequivocally commit to the actions necessary to realize that aspiration. Even so, the Glasgow participants did not reconsider decarbonization, especially not in terms of the costs that would have to be incurred by the developed world to achieve it.

○ ○ ○

Whatever policy the world embraces, the reports of AR6 — even without the overblown rhetoric of commentators — tell us that we must come to terms with certain realities. Again, atmospheric GHG concentrations, principally carbon dioxide ($CO_2$), have been rising steadily since roughly 1850, and are attributable to human consumption of hydrocarbons — initially coal, and subsequently oil and gas. Specifically, the concentration has risen from 280 ppm (parts per million) two hundred years ago to 410 ppm as of 2019. As of 2021, the average global temperature had risen by roughly 1.1°C (since 1850). Rising GHGs in the atmosphere abet the "greenhouse effect." Solar radiation passes through the atmosphere, causing the Earth to warm, while some is emitted back into space. Some gases, most notably $CO_2$ and methane, absorb that radiation and send it back to Earth, resulting in an incremental warming of the Earth's surface. Some other gases, found in aerosols, have the opposite effect, but the net effect has maintained a global average temperature range that enables human existence.

When $CO_2$ is emitted into the atmosphere, whether by anthropogenic or other causes, it does not decompose readily; it persists, on the order of hundreds to thousands of years, depending on aggregate concentrations in the atmosphere. Therefore, as emissions increase, atmospheric concentration increases — assuming humanity makes no efforts to remove $CO_2$ from the atmosphere. Every tonne (a metric ton, or one thousand kilograms) of incremental $CO_2$ increases the potential of incremental global warming and increased concentrations of $CO_2$ in oceans. Meanwhile, about a third of global $CO_2$ emissions are absorbed into the world's oceans, contributing to their increasing acidity, which in turn impacts biodiversity.

The developed world has consumed the greatest amount of hydrocarbon since 1850 and is consequently principally accountable for the rise of GHGs in the atmosphere. The United States is accountable for 20 percent of accumulated GHGs, and Canada is among the top ten countries accountable for accumulated emissions. China is now the largest annual emitter of GHGs, accountable for roughly 25 percent, and India is third, at just below 10 percent. Energy intensity is measured by the quantity of energy required per unit output or activity, so that using less energy to produce a product reduces the intensity. Thanks to technological advances in how we use energy and what forms of energy we use, for developed economies, energy intensity has declined, on average, by about 1.1 percent from 2000 to 2019. This achievement does not, however, alter the fact that accumulated emissions continued to rise over the same period. Global primary energy demand has grown consistently since the 1992 advent of the UN process on climate change; 2020 was an exception, due to the pandemic, and saw demand reduced by 4 percent relative to 2019. Hydrocarbons collectively meet 80 percent of global energy demand, and coal 26 percent. By the end of 2021, demand will likely exceed pre-pandemic levels, and with no material change in the mix of energy to meet that demand.

So, once more, global concentrations of GHG have increased since pre-industrial times due to the consumption of hydrocarbons, and so, in turn, has global average temperature. That is not disputable. What is disputable, however, is the precise functional relationship between these two factors, dubbed "climate sensitivity." That relationship is still intensely debated and far from fully resolved, which makes it fundamentally challenging to forecast global temperature increase.

The UN Intergovernmental Panel on Climate Change aggregates various models developed to project impacts of rising GHG concentrations on global temperature — models that comprise sets of mathematical equations that represent the various physical phenomena that affect global temperature change. These can be integrated with other models designed to project the impact of proposed policies on national economies. The AR6 Working Group 1 report on the status of the physical science on climate change did not generate a specific forecast for global warming; that is, it did not specify the details of a most probable future. Rather, it projected climate effects across a range of scenarios, ranging from no coordinated global policy initiative, equivalent to the existing UN process breaking down, to radical intervention, meaning full realization of decarbonization. That approach provides a credible spectrum of different climate outcomes, measured principally in terms of global average temperature and average sea level. Detractors have assailed the credibility of the models, questioning their ability to replicate past temperature rise and GHG concentrations, and questioning certain variables such as the impact of clouds on the warming process and the specific ranges for climate sensitivity itself. Nevertheless, collectively, these models represent the world's best predictive tool for assessing climate-related policy options.

The AR6 Working Group 1 report describes a global average temperature increase by midcentury ranging from 1.4 to 4.4°C, depending on whether emissions are cut rapidly to net-zero or continue to rise. Containing the increase to 1.5 or 2°C will be impossible unless deep GHG reductions occur almost immediately and will even require ultimately removing $CO_2$ from the atmosphere at a significant scale. We will likely breach 1.5°C in the early 2030s. AR6 concludes that the physical world is expected to respond "proportionate[ly] to the rate of recent temperature change," but that "some aspects may respond disproportionately." This is a somewhat indirect way of saying that temperature change could elicit irreversible or unadaptable catastrophes that genuinely threaten human existence. Cold comfort, perhaps, but the AR6 continues: "However, there is no evidence of such non-linear responses at the global scale in climate projections for the next century, which indicate a near-linear dependence of global temperature on cumulative GHG emissions." This carefully articulated language is meant to convey the conclusion that increased emissions and increased temperature will remain linked, at least for the remainder of this century, in a consistent relationship that when depicted graphically generates a straight line, without any change in slope.

If we accept that the world can adequately minimize climate risk only by containing global temperature increase to no more than 1.5°C above pre-industrial levels, then how much more $CO_2$ can we afford to emit, if any? In other words, what is the global "carbon budget"?

According to the IPCC forecast, the answer is about 450 gigatonnes (a gigatonne is equivalent to one billion tonnes, or 2.2 trillion pounds). Of course, the estimate depends on one accepting the predictive capability of existing climate models. In 2019 humanity emitted over forty gigatonnes of carbon, and even as Covid-19 kept people at home in 2020, global emissions were reduced by only 6 percent. Simple arithmetic suggests that the 1.5°C containment goal will be breached early in the next decade. Furthermore, AR6 predicts that no matter how vigorously the developed world implements decarbonization, we will inevitably overshoot the 1.5°C goal, simply due to the emissions that have already accumulated.

Over time, atmospheric accumulation of GHGs could be contained or reduced by emitting no more incremental GHGs to the atmosphere and also, potentially, by using new technologies to reduce the existing accumulation. As described before, the concept of net-zero emissions, or more to the point, no more incremental emissions, could still accommodate continued use of hydrocarbon, but only if the emissions were captured and sequestered using some technically feasible, environmentally palatable, and affordable technology. The methodology of determining a "carbon budget," however, excludes any of these alternatives as genuinely useful for potentially increasing or stabilizing it.

So, if 1.5°C temperature containment is the accepted global policy objective to deal with climate risk goal, then the world must be constrained to the carbon budget of 450 gigatonnes. Of course, the premise of a global carbon budget immediately raises intractable questions. Who gets what share? Can shares be bought and sold? What are the consequences of non-compliance? Such are the core questions that need addressing if the current UN process is to offer meaningful policy options, though to date the IPCC has hardly confronted them directly.

It is worth noting that the 2015 Paris Climate Accord, which explicitly embraced the 2°C containment target and cited 1.5°C as an aspirational goal, did not allocate any specific portions of the carbon budget to any of the signatory countries. Instead, developed nations were allowed to choose their own targets for emissions reductions between 2015 and 2030. As well, their commitments were merely voluntary, meaning nations would suffer no stated consequences if they did not achieve their targets. There exists no international tribunal to adjudicate on these issues, nor are there tools such as trade sanctions to coerce or encourage adherence. Furthermore, these essentially voluntary commitments taken on by the developed world, even if realized, would, according to the IPCC's models, result in a global average temperature increase of roughly 3°C since the Paris process asked no emission targets of developing countries, including the massive emitters China and India. (Demanding anything more stringent would have threatened to cause the process to break down altogether.) Yale's William Nordhaus, who won the 2020 Nobel Prize winner for his work on climate economics, eloquently and succinctly observed, "Countries have an incentive under Kyoto-type agreements, and all the agreements since then, to talk as if they are going to do something, and then do very little, and ride free on the efforts of others."

○ ○ ○

The question remains whether it would cost the world less to adapt to global warming up to 3°C than to decarbonize enough to keep the change to 1.5°C. But far from being addressed, that question was shunned before decarbonization was embraced by developed economies. Indeed, the idea is utterly dismissed as the equivalent of retrograde climate change denialism because it flies in the face of the accepted orthodoxy, which is enshrined by UN process and ruthlessly advocated by both environmentalists and others sympathetic to fundamental economic and societal change of the existing order. But eventually, the real world will have to deal with the cost of required risk mitigation; and of course, it already began to, throughout the winter 2021–22 energy crisis in Europe.

Conceptually, the real issue is captured in the graph generated by Dr. Nordhaus. We see here the crux of the issue: having no climate policy is not optimal, but neither does it make sense to have a policy objective that is too costly relative to the actual dimensions of the risk itself.

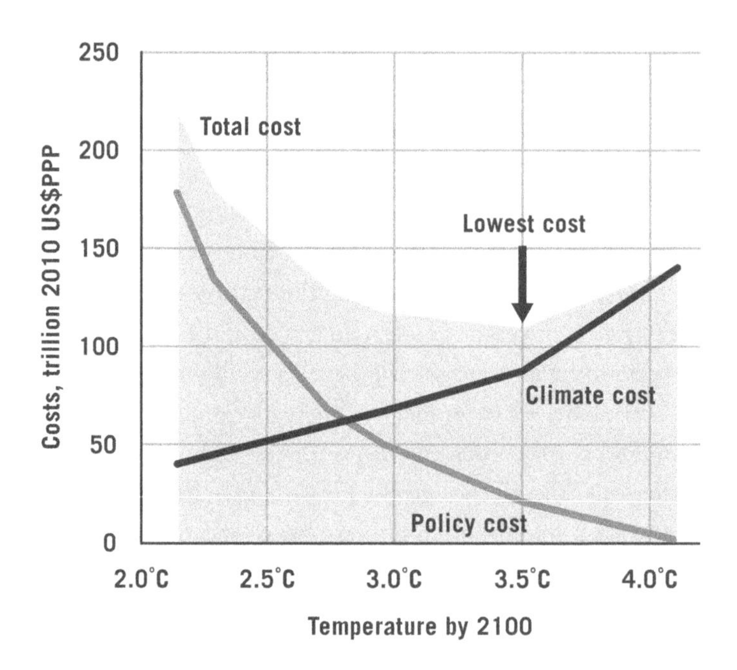


**Conceptually Optimal Climate Policy.** Data from William Nordhaus, "Projections and Uncertainties About Climate Change in an Era of Minimal Climate Policies." *American Economic Journal: Economic Policy* 10, no. 3 (2018): 333–60: doi.org/10.1257/pol.20170046. Graph from Bjørn Lomborg, "Welfare in the 21st Century: Increasing Development, Reducing Inequality, the Impact of Climate Change, and the Cost of Climate Policies." *Technological Forecasting and Social Change* 156 (2020): doi.org/10.1016/j.techfore.2020.119981.

# CHAPTER 3

# Reality: The Grim Cost of Decarbonization

We know that decarbonization is the only route to achieving 1.5°C temperature containment, and the next logical step is to understand how decarbonization could be achieved, in at least some qualitative terms, so that the enormity of the endeavour can be appreciated. Again, it must be stressed that "decarbonization" means no more *incremental* GHGs emitted into the atmosphere — there can be no more GHGs in the atmosphere than are there already. This can be achieved only by generating no more emissions, by capturing and sequestering any further emissions that are generated, or by reducing the existing accumulation of GHGs already in the atmosphere. According to the "carbon budget" logic of the entire UN process, the sooner this process is mandated by world governments, the less likelihood there will be that the world will have to endure excessive years of "overshoot" in excess of 1.5°C before stabilization is achieved.

Set aside the issue of cost versus risk for a moment. Just what does decarbonization physically require for a world that will still use the same amount of primary energy, all things remaining equal? Electricity, to be clear, is not a form of what is called "primary energy," since it depends on a prior, that is, "primary," energy source, hydro, nuclear, solar, wind, or hydrocarbons, to generate it. Today, roughly 83 percent of primary energy is supplied by hydrocarbons — crude oil, natural gas, and coal. Most formulations of decarbonization hold implicit that the same level of economic output can be maintained even with no incremental GHG emissions, but it is also the case that less consumption may be the real goal of many of the advocates of decarbonization, in the environmental movement and beyond. Less consumption in absolute terms would lead to less economic output and inevitably a world with reduced living standards. Obviously, such a world would be less impactful on the physical environment.

First, we must appreciate that hydrocarbons are consumed for purposes far beyond creating heat for industrial and domestic purposes. Hydrocarbons are processed to produce a vast array of petrochemicals that have become essentially indispensable products required in modern life, from plastics to fertilizers to hydrogen. Various sectors would need to be decarbonized, each with its own requirements and challenges.

Consider agriculture. Roughly 20 percent of global GHG emissions impact comes from the agricultural sector of the economy, mostly from cattle emitting methane. Recall that methane is a more effective GHG than $CO_2$ in terms of its capacity to retain heat; however, it does decompose more quickly. If cows were identified as a country for the purposes of attributing global carbon emissions, their annual emissions would compare roughly to current U.S. annual levels, the equivalent of about eight gigatonnes of $CO_2$ in terms of global warming potential. Dealing with this would imply massive changes in diet, especially in terms of how protein is accessed. The basic food staple of rice is produced globally via flooding paddies, giving rise to collateral organic material that again emits substantial methane. There is no apparent production alternative. In Canada, agricultural emissions comprise only about 10 percent of national emissions, reflecting the higher proportion of emissions attributable to hydrocarbon production, transmission, and consumption. Of course, most of Canada's red meat production is based in Alberta, as is most of its hydrocarbon production. Is the consumption of red meat simply to be prohibited or deeply constrained, regardless of how protein will be made available in diets with comparable efficacy and scale? Is vegetarianism a necessary complement to decarbonization?

Road transport — internal combustion and diesel engines using liquid hydrocarbons to provide personal and commercial mobility — is responsible for another 15 percent of global emissions. The apparent alternative is to resort to all electric vehicles running on electricity generated from low-carbon sources, such as renewables or nuclear, and as well ensuring that all the elements of the supply chain that make the components of electric vehicles also use low-carbon electricity, especially in respect of mining critical metal components.

The transition for space heating, accountable for roughly 7 percent of global emissions, would require converting all private residences, commercial spaces, and certain industrial production facilities from hydrocarbon fuels, now mostly natural gas, to electric space and water heating systems (such as resistance heating or heat pumps). Hydrocarbon space heating or cooking applications would be banned going forward. Again, the source of the electricity would have to be low carbon.

Reducing and eliminating emissions from oil and gas production, and reducing further methane emissions in North America from existing oil and gas production and transmission, will require more investment in leak detection and gas recovery. The process will be challenging, technically and economically. But, of course, if aggregate demand for oil and gas reduces, then fewer of these fugitive emissions will occur in lockstep with global production. Electrification of industrial processes for certain energy-intensive industries including steel and cement manufacturing may not be possible, and carbon capture and storage (CCS) will be necessary if hydrocarbons are to still be used to achieve required temperatures.

And then there is the largest contributor to global emissions: the electric generation sector, which is responsible for producing 40 percent of GHGs. That will only increase if these various end-use energy applications electrify, and the actual generation mix of electricity generation does not evolve to low carbon. This implies the end of coal and natural gas as fuels to produce electricity, unless the resultant emissions are captured and sequestered. Renewables, essentially wind, solar, and, potentially, additional hydro, and their associated storage systems, would meet electric demands going forward, in addition to nuclear, assuming that incremental additions of nuclear capacity are deemed acceptable. I emphasize again that the various supply chains required to produce massive incremental renewable capacity must themselves be fuelled by low-carbon electricity. To state the obvious, in the world of net-zero emissions, existing fossil fuels–based generation, currently 60 to 70 percent of total capacity in the United States, would need to implement these conversions, regardless of their economic utility. Implicit in any increased reliance on electricity, and within that, renewables, are massive investments in the transmission grids to ensure that incremental renewable capacity can actually reach existing demand centres. Furthermore, incremental grid investments will be required to ensure real-time reliability despite fluctuations in wind and solar availability.

It bears emphasizing that much of the environmental movement rejects the option of nuclear energy, notwithstanding that it is non-emissive and not subject to the intermittency that bedevils wind and solar. The movement's extreme resistance to nuclear relies on claims that the inherent risks of such operations, and the long-term containment of spent nuclear fuel, are unjustifiable. This despite the long record of reliable and safe operations of nuclear power stations from Ontario to France to the southern United States. Bill Gates, if not John Kerry, has conceded that decarbonization is difficult to conceive without massive reliance on nuclear power. "If we're serious about solving climate change," Gates stated, "and quite frankly we have to be, the first thing we should do is keep safe reactors operating.... [But] even then, just maintaining that status quo is not enough. We need more nuclear power to zero out emissions in America and to prevent a climate disaster."

Marine transport and aviation require a longer-term conversion to liquid fuels based the conversion of biomass, most promisingly algae. This is likely to be the most difficult substitution for diesel and jet fuel derived from crude oil. Even the contention that biomass itself can be carbon neutral remains an open question. As for the manufacture of ethylene, the key building block of modern petrochemicals and other refinery products, CCS systems will have to be implemented to minimize emissions from initial chemical processing and related heat requirements.

Cement manufacture is also problematic: emissions occur in part because of the heat required for the chemical process to occur, and then subsequently when the chemical reaction itself produces the calcium carbonate. Whether CCS methods can be applied economically to cement manufacture remains uncertain. Alternatives to the existing chemical process are not yet really at hand, and the same applies to steel manufacturing, which requires metallurgical coal.

As can be seen from the above litany, where substituting hydrocarbons with electricity from non-emissive sources cannot work, effecting decarbonization will require the application of CCS. But there are very few commercial uses for captured $CO_2$; the value of CCS is really only to avoid carbon emissions. Capturing $CO_2$ from post-combustion flue gas streams (essentially, what goes up a smokestack) can be accomplished technically, but the cost, in even "best-of" applications, is significant.

The best-of applications in this context are those in which the concentration of $CO_2$ is relatively high in the effluent or already at relatively high pressure. Examples include flue stacks from coal generation facilities, gas reforming units, and hydrocarbon gasification units. Even for those, the cost is high.[1] Some applications contain less carbon to begin with, such as natural gas. The cost of CCS is even higher in these cases because the stream is more dilute and greater compression is required to effect the separation and transmission to sequestration reservoirs. Beyond the separation step, the captured $CO_2$ must still be pipelined and sequestered. These economics also depend on the proximity of adequate sequestration reservoirs and regulatory constraints. And finally, the impact of scale on the separation required makes the application of CCS more problematic for smaller, remote applications such as steam injection–based oil recovery operations.

Lastly, any decarbonization scenario that relies substantially on wind and solar requires electricity storage. It is axiomatic that natural gas–based generation would not be available to deal with intermittency in a fully decarbonized future. Apart from the not-immaterial geopolitical and economic considerations of supplying required minerals for the manufacture of utility-scale battery systems, can such systems reliably deliver back power not simply for portions of hours, but on the order of days? Even in California, which has actually, and incredibly, already legislated that it will by 2045 have an electric generation system based solely on renewables and battery storage, those accountable for trying to effect this conversion acknowledge that adequate battery storage remains highly problematic.

Increased electrical requirements will require more electric transmission, inevitably in the form of more high-voltage above-ground transmission systems, and likely across substantial distances since demand and supply are often far apart. The same resistance that has impacted hydrocarbon pipeline development will also impact electric transmission infrastructure, especially in those jurisdictions between the source of renewable power and the final consumption area. A classic recent example occurred in Maine in November 2021, when voters rejected by 61 percent a US$1 billion high-voltage transmission project supplying surplus hydro power from Quebec to Massachusetts. This kind of response will doubtless be replicated in other jurisdictions if similar projects are attempted. And to date, the voters of Maine have the legal right to resist, regardless of the imperative to decarbonization, and regardless of specific benefits of the project that might accrue to the state.[2]

Beyond electricity, hydrogen comprises a potential element of any decarbonized future, because burning it produces no emissions; however, making it to scale, affordably, remains elusive. Today, hydrogen is produced and used to upgrade certain hydrocarbon feedstocks, notably Alberta bitumen. It is made by reforming natural gas; that is, by separating the hydrogen from the methane molecule, generating $CO_2$ as a by-product. Electrolytic processes can separate hydrogen from water, but they are not applied industrially. Though it also presents challenges related to containment and metallurgy, hydrogen is still held out by some as a "Holy Grail" of energy transition.

Some take solace in the possible pathways to decarbonization based on substantially existing or developing technology. But these potential technologies are problematic in terms of absolute cost, normalized for reliability relative to utilizing hydrocarbons, in whole or in part. Meanwhile, most attempts at describing some actual pathway to stabilizing at 1.5 or 2.0°C ignore or diminish the potential of $CO_2$ removal (CDR) — literally, carbon dioxide removal — also known as negative emissions, which encompasses a range of techniques to take $CO_2$ from the atmosphere and store it in "geological, terrestrial or ocean reservoirs, or in products." Some of these are "nature-based solutions," most notably reforestation, and some involve more industrial processes, such as direct air capture and bioenergy with carbon capture and storage (BECCS). One would rationally expect that these could be applied to offset emissions from those sectors that are too difficult, that is, too costly and too valuable, to decarbonize, such as petrochemicals and aviation. Yet it is rarely conceded, let alone emphasized, that CDR could also be implemented at a large scale to generate global net-negative $CO_2$ emissions, resulting in anthropogenic $CO_2$ removals exceeding anthropogenic emissions, to thereby meet some long-term climate stabilization goal other than simple decarbonization.

For many in the environmental community, the entire prospect of CDR technology evolving to the point where it does materially stabilize atmospheric concentrations of GHGs represents an untenable moral hazard, where any momentum for mitigating GHG emissions is reduced and serves to reinforce hydrocarbon consumption. Of course, if such technologies were to be perfected at scale, inclusive of all relevant economic and environmental considerations, rather than a moral hazard it would be a potential panacea or, more likely, a major adjunct to adaptation investments. One would hope an open-mindedness to such technology development would apply.

For Canada specifically, it always bears repeating that most of the country's electric generation capacity is already low carbon, based on decisions made long before Canada was given any credit for those decisions. Less than 15 percent is fossil fuels based, and, again, most of that is located in Alberta. Future Canadian emissions reductions could come from Canada contracting, if not outright deconstructing, its hydrocarbon production industry, assuming that Canada could afford the value loss and that no other country would simply pick up Canada's market share. But this is a brutal choice for Canadians if decarbonization is to be taken at all seriously.

Decarbonization ultimately comes down to electrification, CCS, and potentially the use of biomass and hydrogen as replacement fuels. All are problematic and imperfect. All of these are touted for their ability to help achieve GHG mitigation, that is, their utility in eliminating the activities that cause emissions. Those who promote these never speak of adaptation, but it will occur in many forms if markets are allowed to internalize the cost of accommodating to climate change appropriately. Adaptation reduces the cost of global warming, but it has been seen by most of the environmental movement as at best a complement to the decarbonization objective, never as a real potential policy alternative.

Adaptation involves learning to live with climate change, changing lifestyles and activities in order to reduce its harm, while mitigation requires a reduction of the risk of climate change by eliminating activities that cause emissions. Classic examples of adaptation include irrigation of crops where higher temperatures and reduced rainfall cause drier soils; migration to cooler countries; better forest management — essentially, clearing dead wood to reduce the risk of forest fires — and reforestation; increased use of air conditioning; atmospheric removal of GHGs; urban densification, which leads to less use of personal vehicles and greater energy efficiency from centralized heating and cooling systems; and weather forecasting, which enables more proactive personal and community responses to extreme weather events.

The most significant example of mitigation is the substituting of hydrocarbon-fuelled electric generation with non-emissive sources of electricity, either nuclear or renewables. Other notable examples of mitigation include eliminating any appliance or vehicle that runs on hydrocarbon fuel and using replacements that use electricity made from a non-emissive source.

The distinction between adaptation and mitigation can sometimes become blurred, such as when the cost of emitting rises due to carbon pricing and evokes market responses, such as retrofitting houses. In that case, fewer emissions are generated — mitigation — but the capital stock (the inventory of physical infrastructure that enables the production of goods and services) also changes, adapting how the basic need for shelter is provided.

On a global basis, what would it cost to achieve net-zero emissions, or put more honestly, decarbonization? According to the Energy Transition Commission, the global coalition of largely private-sector interests committed to this goal, the answer is $1 trillion to $2 trillion per year of additional investments, or 1 to 1.5 percent of global GDP, until midcentury. At the end of 2021, the UN AR6 had not released its analysis of the economic considerations of net-zero emissions, only the physical implications of climate change via the Working Group 1 Report released in August 2021. In January 2022, McKinsey Consulting estimated that the project will cost $9.2 trillion annually until 2050. That is $3.5 trillion more per year than the world is currently laying out for both low-carbon and fossil fuel infrastructure. For context, global GDP for 2019 was on the order of $87 trillion. Covid-19 set back the world economy by about 4.4 percent for 2020, and that created a 6.6 percent drop in emissions. The arithmetic is clear: roughly a tenth of global GDP would be required to meet the decarbonization objective. The McKinsey authors concede, with crushing irony, "Reaching net-zero emissions will require a transformation of the global economy."

Another way to make the cost of decarbonization more tangible is to calculate the carbon tax required to make the transition economic, or even to achieve substantial progress toward decarbonization. In recent years, few public reports have been made public on actual abatement curves for the various decarbonization alternatives, and the McKinsey report contains none. We need an abatement curve defining how high a carbon tax must be imposed to create an economic incentive to reduce a specific GHG emission; in other words, we need to know the cost of alternative technology to eliminate a specific GHG emission. If such curves can be assembled, they can guide governments to set the precise carbon taxes required to generate any given level of emissions reductions. This assumes, it must be noted, that all emitters react in an economically rational manner to the level of carbon tax imposed.

Canada will impose a CDN$170/tonne carbon tax on its emissions by 2030, a pricing level as high as any developed economy has yet imposed on itself, including the countries in the European Union (EU). But how much abatement does that buy? The federal government has never explicitly answered that question. Specific engineering calculations would have to be carried out to determine the specific level of carbon tax required to make each emissions reduction opportunity economic. The abatement cost necessary to achieve a specific objective of emissions reduction would set the price for all emissions. The greater the level of emissions reduction required, the higher the required tax. If absolute decarbonization is the actual objective, the tax would be very high, essentially infinite, the equivalent to prohibiting of hydrocarbon utilization. A carbon tax of US$200/tonne would create a real incentive for lower-carbon alternatives, especially in the electric generation sector, but it would be unlikely to validate all CCS applications economically or to impact demand for petrochemicals significantly.

Could it be that net-zero emissions, if actually pursued by the world's developed economies, will make the world poorer than just living with the risk? This is a legitimate question. The AR5 IPPC synthesis report was ambiguous on this point, at least in terms of providing a transparent comparison of cost and benefits. One hopes the 2022 AR6 is more useful.

The unconfronted question remains whether decarbonization is the optimal global climate policy. Do other levels of mitigation and adaptation give sufficient risk reduction at less cost for economies? Is there a limit to how high a carbon tax applied on emissions should be before the costs of abatement are greater than the risks being reduced? Living with some climate change risk and dealing with the consequences, the extent and timing of which remain uncertain, may be more optimal in an economic sense. Or, put more bluntly, does a 3°C increase in average global temperatures comprise an existential threat to humanity?

The social cost of carbon should have driven the UN process, but that has never been the case. And what is the social cost of carbon, precisely? Put as simply as possible: it is net cost created by an incremental carbon emission. Climate change due to incremental carbon emissions from hydrocarbon consumption create costs as already described, for example, impacts on agricultural productivity, human health, rising sea levels, floods, wildfires, and so on. But concurrently, mitigating or adapting to those impacts has a cost. And using hydrocarbons creates value. To the extent that costs exceed benefits, that economic impact should be taken into account when carbon is emitted, and emitters should pay the net cost. This would create the foundation of a logical carbon pricing system that allows the full cost of using hydrocarbons to be internalized by the global economy. Assuming that the carbon price is applied via a carbon tax on emissions, part of the funds generated could be invested directly in adaptation, research on improving low-carbon alternatives economically, and existing mitigation alternatives in certain sectors. And, of course, the carbon tax would also influence various elements of the economy, from businesses to households, to alter their operations and investments, leading to reduced emissions. The key point is that the amount of emissions reduction is not specified directly. It occurs as a function of the price level reflected in the carbon tax, which should ideally equal the social cost of carbon. Ultimately, all nations of the world would apply the same carbon tax.

Who could argue with the elegance and logic of carbon pricing based on the social cost of carbon? Let resulting markets sort out what substitution and mitigation technologies make economic sense, what level of reduced consumption of energy as hydrocarbons is justified, and what adaptation infrastructure can be justified economically. Even though climate economists fully subscribe to this construct, the UN process has never made carbon pricing its focus. The environmental movement has rejected it, preferring command-and-control regulation. Classic examples of such regulation include explicit limits on emissions in specific sectors, such as the emissions cap in Canada that prohibits using natural gas for households. California and British Columbia have declared that their residents, after a certain date, will not be allowed to acquire new vehicles that use internal combustion engines and have mandated phasing out coal and natural gas for electric generation. And even in jurisdictions where some form of carbon pricing is applied, most notably California and the EU, the actual pricing has not been premised on rising over time to the only logical economic upper limit, the "social cost of carbon." Rather, the price has been set to achieve some specific level of emissions reduction based on other risk mitigation criteria. This choice will, in the end, doom the policy. The social cost of carbon, which comprises an estimate of the net economic costs of emitting one additional tonne of the carbon dioxide into the atmosphere, must be part of any rational policy if it is to be effective. What is key here is the term *net*. It is essential to consider the net of the benefits derived from that emission — the price that we pay to generate an emissions reduction should not be more costly than the benefits that accrue from utilizing carbon fuels. If it is, the reductions cannot be justified.

In practice, the carbon pricing that was imposed remained at levels well below what could realize the Paris Accord targets and did not even reach some deemed social cost of carbon. This is perhaps unsurprising. As Roger Pielke, a political scientist at the University of Colorado known for his work on climate policy, states, there exists an "iron law of climate policy." According to him, "neither the public nor politicians will accept economic contraction for the purpose of reducing carbon emissions." History lends credibility to his formulation.

Admittedly, carrying out the calculation is difficult, because it requires integrating economic models with those relating to the physical impacts of carbon emissions, and beyond the complexities implicit in that integration, certain contentious assumptions must be made about the value of human life and the value of near term impacts relative to future ones. Although economics has a fully evolved means of dealing with those issues, they remain intensely debated and disputed across the political spectrum. A human life has a value above zero, but it does not have infinite value, at least not economically. In the real world, discount rates are fundamental to all financial analysis, as capital has to be paid for, and the time value of money is real: cash flows earned sooner are more valuable than those earned in the distant future. Anyone with an interest-bearing bank account knows that costs incurred in the future are not the same as costs incurred today, and the same is true of benefits. John Maynard Keynes's infamous quip "In the long run, we are all dead" makes the same point.

These two considerations, more than any others, have kept the social cost of carbon from ever being calculated credibly at levels that would drive decarbonization meaningfully. The entire thrust of the UN process should recognize that, but never has. The Obama administration's interagency working group tasked to solve for the social cost of carbon found a median value around $50/tonne for emitted carbon, with an upper range of $100/tonne. Perhaps not surprisingly, the Trump administration declared the cost lower by almost an order of magnitude, consistent with its members' climate change denialism orientation. Sadly, the Trump administration might have been well advised to accept $50/tonne and go on to use that benchmark to reinvent the UN climate change process; that, obviously, did not occur.

The two vital points here are, first, that achieving decarbonization is expensive, perhaps infinitely expensive, based on the status of technologies currently available. Second, and intractably, replicating the "gift" that hydrocarbons represent is immensely difficult. Hydrocarbons offer massive energy stored via fundamental chemical properties, available when needed at minimal cost. The substitutes are intermittent and inherently more costly, due to the complexity of their production. Moreover, they often rely on mineral components that are themselves costly and energy intensive to produce. Again, the cost of GHG emissions have to be considered in an overall context, namely the social cost of carbon.

○ ○ ○

In 2018, just as the IPCC issued its special report calling for 1.5°C temperature containment and decarbonization, Dr. William Nordhaus of Yale University won the Nobel Prize for Economics, specifically climate economics. His modelling projected that the "optimal" amount of global warming by the year 2100 would be 3.5°C, a full two degrees higher than the 1.5°C containment diktat of the IPCC. His work led logically to the conclusion that 1.5°C containment would harm the global economy more than doing nothing at all about climate change. Put even more bluntly, 1.5°C containment is ludicrously expensive given the existing available substitution and mitigation technologies. Decarbonization is simply too costly for the risk that the world actually faces. The world would be better off living with some risk of extreme weather events and less probable high-impact discontinuities in global climate. Sadly, of course, the UN climate process has not come to that conclusion, but quite the opposite. And even more sadly, post-Trump, none of the political leadership of any developed economy has explicitly resisted 1.5°C containment.

# CHAPTER 4

# Before the Fall: Canadian Energy Developments from the Beginning of 2019 to March 2020

In the years leading up to the advent of Covid-19, the UN climate change process evolved, and the Trudeau government modified its national climate policy to accommodate that evolution. Specifically, it declared its intention to realize the emissions reduction targets coming out of Paris. Over the same period, various major energy infrastructure projects faced obstruction in Canada. It is essential to understand this history because it contextualizes how the objective of decarbonization has already resulted in Canada suffering substantial economic losses that the country has shown no capacity to replace. Canadian and American governments and courts have facilitated that value destruction. In Canada, the same political leadership has persisted into 2022.

Canada's hydrocarbon production sector has been traditionally, and continues to be, a substantial component of the overall Canadian economy. Its primary energy production sector is responsible for about 10 percent of the total Canadian GDP as of 2019. The oil and gas industries, including extraction and support activities, petroleum refining, pipeline transportation, and natural gas distribution, accounted for 7.5 percent of Canada's GDP. Electricity was the second largest energy sector, contributing 1.8 percent of Canada's GDP, and 20 percent of that was still generated by hydrocarbons.

Canada is the world's fourth largest exporter of crude oil, a position it has held since 2015, after the build-up of oil sands production capacity. It is also the world's seventh largest exporter of natural gas, although its ranking has declined from a decade ago when it was ranked third, reflecting the build-up of fracking-based production capacity in the United States and the failure of Canada to seize LNG export opportunities earlier in the 2010s. Energy products have been Canada's largest domestic merchandise export every year since 2010. In 2019, the total net exports (the value of exports minus the value of imports) of energy products totalled CDN$187 billion. In 2018, the energy sector paid $8.2 billion in corporate and indirect taxes, which accounted for 7.1 percent of total corporate and indirect taxes paid to governments. Despite its value to the national economy, the energy sector has, for more than a decade, faced an ever-increasing number of barriers that have limited its ability to operate successfully.[3]

Beginning in 2009, Alberta confronted a long period of dysfunction and frustration in moving its hydrocarbon potential to market — a struggle I witnessed directly in my executive role at TransCanada Pipelines. After my retirement in 2015, I became a commentator on Canadian climate and energy issues, and in my books *Dysfunction* and *Breakdown* I extensively explore this history and its implications for Canada and Alberta. By 2019 and early 2020, the situation had changed substantially for the better, such that a real breakthrough on market access seemed at hand, despite the unfolding convergence of various global forces toward a policy of decarbonizing the developed world's fundamental energy systems as the only way to contain the global climate change risk at acceptable levels.

## Keystone XL

Keystone XL (KXL) was always an iconic project. For the environmental movement of North America, this pipeline threatened, like no other, to enable expansion — on the order of 20 percent of existing Canadian crude oil production — of the production and consumption of one of the world's most carbon-intensive sources of crude oil. It was understood by the environmental movement that if it could not ultimately disable this project, it might not be able to face down any other major hydrocarbon infrastructure project. The tools of resistance were clear: persistent obstruction. The goal: regulatory or political denial by the centre-left Obama administration and Democratic Party. As environmental polemicist Bill McKibben so floridly expressed it, "in deciding to block Keystone, the president would finally signal a shift in policy that matters, finally acknowledge that we have to keep most of the carbon that's still in the ground in that ground if we want our children and grandchildren to live on a planet worth inhabiting."

Conversely, for the North American hydrocarbon production industry and virtually all of its supporters, the project amounted to conventional pipeline technology applied with maximum economy and long-term security to meet a legitimate market demand. After all, production and consumption were hardly illegal. The value enabled for Canada, and Alberta especially, promised to be substantial and sustainable well into the middle of the century.

KXL was proposed as an extension of the existing Keystone system, which had been operating since 2010, following its approval by George W. Bush. Keystone transported Alberta diluted bitumen (dilbit) to mid-continent markets, and the new expansion would carry one million barrels of dilbit per day from the oil sands to the U.S. Gulf Coast refinery complex, which was optimized to run on heavy oils and constituted the product's most lucrative end market. Certain American producers, notably in North Dakota, would use the pipeline, too, to a lesser extent. Completing KXL would require a capital investment exceeding $10 billion.

Diluted bitumen from Alberta's oil sands has a carbon intensity (the ratio of attributable incremental emissions per barrel produced) comparable with that of other heavy oils produced globally, most saliently Mexican and Venezuelan oil, its most direct competitors. However, it does have a carbon intensity significantly higher than that of other crude oil production that does not require significant energy input or subsequent upgrading. Despite that fact, it must be stressed that global demand for such heavy oil would persist, whether or not Alberta dilbit satisfied it. Consequently, the advent of KXL would not materially impact global emissions, nor would the Alberta dilbit carried by it. Nor, for that matter, would the associated American emissions. In other words, no material incremental emissions were attributable to Alberta diluted bitumen.

Furthermore, KXL's proposed direct diagonal route through the middle of the continent, from southwest Alberta through Montana, South Dakota, and Nebraska, would, with its capacity of up to one million bbl/day, provide transportation economics superior to those of other available pipelines and alternative modes of transportation — potentially on the order of US$7–8/bbl relative to rail or a more circuitous pipeline option, if such options were even available. That could represent up to $3 billion annually in improved cash flow just on improved transportation economics. And that value would be substantially higher, likely an order of magnitude higher, if KXL enabled true incremental production. This reality was validated by major credit-worthy production and refining interests that supported the project commercially with long-term shipping commitments. Further substantial economic benefits would accrue to other sectors over the project's construction phase, both in Canada and the United States. And what is $30 billion of incremental value to Canada? For context, the Canadian federal deficit was $365 billion in 2019.

The first of KXL's American regulatory assessments, completed in 2011, demanded special operations conditions to reduce the possibility of oil spills and other operating disruptions, and the project incorporated these. Meanwhile, the limited portions of KXL within Canada had been approved back in 2010 by the National Energy Board, with Cabinet approval. It was understood that KXL would cause carbon emissions generated within Canada to increase as oil sands production exceeded three million barrels per day by the end of 2019, potentially rising over time to up to four million.

Emissions attributed to Canada would remain in the order of seven hundred to eight hundred megatonnes annually, less than 2 percent of global emissions. Of course, these numbers were problematic, considering the emissions reduction commitments Canada made at the Paris Climate Conference, to reduce national emissions to 30 percent of 2005 levels, or roughly 200 to 250 megatonnes per year. Still, on a global basis, KXL would make no difference to the global demand for heavy oil, nor to the global emissions created by meeting that demand. Whether Canada produced the oil or some other country did, the same demand would be met. The only question was what country derived the value and had the related emissions assigned to it by the UN process.

Any objective review of the 2011 or 2014 regulatory assessments undertaken by the Obama administration could only construe them as validating the substantial net economic benefits of the project to both Canada and the United States. The pipeline's incremental emissions impact globally was deemed immaterial, and the assessment found no other environmental "showstoppers" that had not or could have been addressed by the various conditions already imposed on the project's operations and construction.

Despite this regulatory record, built up over almost eight years of review across his two administrations, in November 2015, just before the UN's climate conference in Paris, Obama announced that he would not provide the necessary presidential permit to allow the pipeline to cross the Canada-U.S. border. In other words, he rejected the project. Providing the permit, would, Obama rationalized, impair his "climate bona fides" and disable his ability to provide U.S. leadership at the upcoming Paris Climate Conference. "America is now a global leader when it comes to taking serious action to fight climate change," Obama said in remarks from the White House. "And, frankly, approving this project would have undercut that global leadership." This, after his own 2013 statement, at a press conference about climate change in Georgetown, that his disposition on KXL would be based on the "materiality" of its impact on global emissions. When Obama rejected KXL at the end of November 2015, he made no attempt to apply that standard, nor even to rationalize why it had been ignored. Most audaciously, he ignored his own administration's environment assessment, published in early 2014, which represented a robust foundation for approval.

The only section of pipe that Obama's administration ever facilitated was that between Nebraska and the Gulf Coast, constructed in 2012. This construction went forward partly because its commercial justification did not depend directly on the cross-border permit, and incremental construction activity looked good for Obama's 2012 re-election. Otherwise, the regulatory process, along with related lobbying and advocacy, proved a forlorn and disingenuous process, dominated by politics — politics over the integrity of the regulatory process; over relations with Canada, a major trading partner; and over tangible economic value — and without nuance. Environmental activists within the Democratic Party, and, more importantly, donors to the Democratic Party, simply insisted that permission to build the rest of KXL should be denied. No Democratic politician, besides a few hanging on in the U.S. Senate from Louisiana and West Virginia, could afford to ignore that demand.

That rejection came down under the watch of Canada's newly elected Liberal government, led by Prime Minister Justin Trudeau. Trudeau accepted Obama's decision pliantly, without expressing any real anger, and without acknowledging the real economic costs that had been imposed on Canada. Trudeau's acquiescence alienated not only industry, but most Albertans. He was clearly not a champion of their interests and did not even seem to appreciate what had been lost to the country. "We are disappointed by the decision," he stated, "but respect the right of the United States to make the decision. The Canada-U.S. relationship is much bigger than any one project, and I look forward to a fresh start with President Obama to strengthen our remarkable ties in a spirit of friendship and co-operation."

When it came to climate, Trudeau thus revealed that his fundamental empathies and political instincts in respect of climate and energy were no different from Obama's. He would abide hydrocarbon production only grudgingly, regardless of its economic value, and avoid seeking any real balance between climate risk mitigation and energy affordability. During an Ontario town hall, Trudeau stated, "We can't shut down the oil sands tomorrow. We need to phase them out. We need to manage the transition off our dependence on fossil fuels." The environmental movement framed lowering emissions as nothing less than a moral imperative, and Trudeau would not contradict that position by weighing the cost of hydrocarbons against their benefits.

Losing KXL hit the industry and the province of Alberta particularly hard because oil pricing had adjusted downward to the range of $50/bbl after fifteen years in the $80/bbl to $100/bbl range. The halcyon days were over when oil sands investment economics had been at their most compelling, and billions were invested in Alberta. What Alberta had lost on price, it had sought to recover with increased volume, but that strategy depended on infrastructure such as KXL to accommodate the growth.

In 2016 TransCanada and its industry partners responded to the cancellation of the pipeline by suing the Obama administration for damages. The claim, on the order of US$4 billion, rested on the contention that Obama's decision to deny the KXL's presidential permit was at odds with the regulatory record that had been developed for the project.

Before the launch of the suit, the prime minister underscored where his sympathies lay when, at the Paris Climate Conference at the end of 2015, he signed on to commitments as ambitious as Obama's, reversing the equivocations and treaty withdrawals of his predecessor, Stephen Harper. Trudeau made no attempt to explain how Canada would achieve those commitments, nor did his government seem to have seriously considered the question. Not only did Trudeau clarify no plan to meet Canada's new emissions reduction commitments, but those commitments fundamentally contradicted any increased oil sands production and liquid natural gas (LNG) export from Alberta. Trudeau knew the province fully expected to grow its LNG industry, and even Alberta's NDP premier, Rachel Notley, with her fervent environmentalist convictions, could not ignore the basic fiscal realities of the province she led. Meeting Canada's Paris commitments and accommodating Alberta's growth were fundamentally contradictory policy objectives, and all but impossible to reconcile. Did dealing responsibly with climate change risk really amount to eliminating all hydrocarbon growth in Alberta?

Not only did the Trudeau government accept the loss of the KXL pipeline, but it also killed two domestic alternatives, Northern Gateway and

Energy East, deeming them too problematic to approve. Northern Gateway represented a direct route from Edmonton, Alberta's crude oil collection hub, to the Pacific Coast — it essentially went straight west to Kitimat, British Columbia, thereby creating the potential for more market diversity via access to tidewater. The route obviously avoided the Lower Mainland geography and demography.

Trudeau and his principal adviser, Gerald Butts, the *bête noire* of Canadian hydrocarbon interests, had a long-standing antipathy to any crude export development in the Kitimat area, notwithstanding the determination of the federal regulator that constructing the project and putting it into operation were in the public interest. Butts made his mindset clear: "From hewers of wood and drawers of water to makers of moonscapes and creators of toxic tailings ponds: what a face for Canada to show the world." He called for "a moratorium on expansion of tar sands development and halt [to] further approval of infrastructure that would lock us into using dirty liquid fuels."

For Energy East, a project based on converted underutilized natural gas capacity between Alberta and Ontario and constructing new pipe across Quebec into New Brunswick for tidewater access, Trudeau was unwilling to confront NIMBY resistance in Quebec by asserting federal jurisdiction. This translated into untenable extension of the regulatory process, to the point that the developer, TransCanada, could no longer trust the good faith of the Trudeau government. In words instructive as they were depressing, he stated, "Governments grant permits, but communities grant permission."[4] He never explained how that could be compatible with the due process of Canada's regulatory approval system, which bases its decisions on legitimate cost/benefit analysis, nor did he elaborate on which "communities" he meant. Which entities would receive de facto vetoes? Trudeau's position illustrated his contempt for and ignorance of the necessities of capital investment.

The Trans Mountain expansion pipeline (TMX), an expansion of an existing pipeline between Edmonton and Burnaby, British Columbia, with a scale of up to 800,000 bbl/day of incremental capacity, was deemed by the federal government to be the least objectionable alternative for transporting oil sands dilbit to market, and remained on the table. It was conceived in the same era as KXL, as an expansion of the existing Trans Mountain system, historically intended to move Alberta crude to refining facilities in B.C.'s lower mainland. Its capacity would double both the basic trunkline and terminalling facilities in order to provide increased export capacity. From there, a hand-off to tankers would provide access to diverse markets, including U.S. Gulf Coast refinery markets via the Panama Canal. Trudeau and Notley struck an uneasy compromise: TMX would be allowed, and Alberta would agree to a limit on its oil sands emissions, introducing a production cap of approximately four million barrels per day — a limit that would remain if no fundamental carbon-mitigation technology breakthroughs arrived. As a quid pro quo, Alberta would accept Canada's Paris Accord emissions reduction targets and impose on itself carbon pricing and other undertakings, particularly in how the province generated its electricity.

Did Trudeau accept TMX in late 2015 because he recognized that oil sands production represented real economic value to Canada? Did he see that eliminating oils sands production in Canada would not eliminate heavy oil demand and the emissions related to their production? Was his decision based on any cost/benefit analysis of incremental cash flow to Canada from increased oils sands production relative to any quantified climate impact directly attributed to it, whether in terms of local air and water quality or the contribution of incremental GHG emissions to global temperature increase? He never said as much, nor provided pertinent analyses.

As Trudeau and Notley found their compromise, they, like everyone else, assumed that, across the border, Democrats would remain in power — specifically that Hillary Clinton would win the 2016 U.S. election — and that Canada would collaborate with the United States on a continental approach to reducing emissions and at least progressing toward meeting both nations' targets. Of course, Donald Trump's election in November 2016 changed everything. Virtually overnight, KXL was revived as a serious project.

Trump had made clear that he did not take climate change risk seriously, so he had no interest in halting energy projects that provided economic benefits. His position affected not only KXL, but it had profound implications for American energy policy generally. Trump withdrew the United States from the UN process on climate and cancelled its specific emissions reduction undertakings. Under Trump, the U.S. executive branch essentially embraced climate change risk denialism, and as an obvious corollary enabled unrestrained continental hydrocarbon development, primarily by removing regulatory obstruction. With permission for the construction of its pipeline back in place, TransCanada had to reassemble the commercial support that had underpinned KXL from 2007 to 2015. It also needed to reassure itself that the pipeline could be built before Trump's presidency ended. The possibility of obstruction from the environment movement and resort to an empathetic Democratic-appointed judiciary were inevitable risks. Even worse, if Trump were not re-elected there was the real possibility that a future Democratic administration could revoke the presidential permit, regardless of how much had been invested in the revived project up to that point.

In May 2017, the Trump administration issued TransCanada its long-sought presidential permit for KXL to cross Canada-U.S. border, relying on the 2014 environmental assessment carried out by the Department of State under Obama. Trump signed the permit, handed it to TransCanada CEO Russ Girling, and glibly inquired when he could expect to see "bulldozers in the field." Girling's response was more cautious and reserved than Trump expected. The project still needed to confirm its route through Nebraska, where activists had long been working to obstruct it. (Admittedly, the resistance to KXL in Nebraska was driven substantially by the route the pipeline would take, not the nexus between KXL and global temperature increase.) The fight could still take several months, Girling admitted.

The issue of the route through Nebraska had been contentious from the outset of the project, contributing delays and uncertainties into the approval process even in 2019. In fact, Obama had relied on the Nebraska controversy back in 2011, when he forced TransCanada to recommence the entire approval process after nearly completing it. Specifically, he required a new route that avoided directly impacting the Nebraska Sandhills. The region of mixed-grass prairie on grass-stabilized sand dunes covers just over one-quarter of the state and had been deemed uniquely sensitive. TransCanada responded to Obama's demand with a new, more circuitous, route, which the Nebraska regulatory authority approved in 2015. However, various Nebraskan environmental and land-use groups contested that approval, and when Trump provided his permit in 2017, the decision remained before the Nebraska Supreme Court. The issue was whether the state regulator had the authority to modify TransCanada's revised route as part of its approval, and, incredibly, the decision didn't come down until August 2019 — finally validating that KXL had a legally secure, albeit more costly, route through Nebraska.

As 2017 progressed, the leading American environmental non-governmental organizations (ENGOs) launched further litigation against the project, arguing that Trump's permit was predicated on a 2014 environment assessment no longer up to date. In January 2018, while this litigation was still being played out, TransCanada announced that it had acquired sufficient shipper support for the revised project. Incredibly, after more than a decade of obstruction and market evolution, the project's underlying commercial premises were still intact: a direct pipeline nexus between Alberta bitumen and Gulf Coast heavy oil refining capacity made financial sense.

But even with all this in place, it was not until late 2019 that TC Energy (the recently renamed TransCanada Pipelines) found itself poised to make a final investment decision to proceed. The company did so despite the 2018 decision of an Obama appointee on the District Court of Montana to vacate the 2017 cross-border approval pending a supplemental environmental review. This review was meant to show whether the 2014 assessment adequately justified Obama's 2015 denial. TC Energy and the Trump administration did provide some amendments to the 2014 environment assessment in March 2019, and Trump issued another cross-border permit — directly this time, not via the Department of State, as had been the case previously for both Keystone and KXL. This presidential sole discretion made any binding judicial review impossible. In mid-July 2019, the U.S. 9th Circuit Court of Appeal affirmed that decision.

From November 2016 until the end of 2019, TransCanada's management was extremely cautious in its approach to developing the pipeline, anxious to avoid making the same mistake as the first time around. That is, they avoided ramping up spending, particularly making capital equipment commitments in pipe and pumps, until they had dealt with all the legal obstructions and had all the requisite approvals in hand. TransCanada's first attempt at getting KXL approved resulted in sunk costs of more than $4 billion; the company's claims against the Obama administration were, in large part, an attempt to recover those costs. Proceeding more cautiously this time would, of course, compress the time available to complete the project. A Trump re-election was no better than a fifty-fifty prospect by the end of 2019, so proceeding with the project meant taking on the risk of significant further spending, all with the knowledge that a subsequent Democratic administration could disable the project again, even after significant spending had been incurred.

In March 2020, TC Energy finally announced that the revised KXL would move on to construction. Almost exactly three years had passed since the euphoria of November 6, 2016, when Trump finally issued that presidential permit, and getting to this point had required Alberta's government to intervene, becoming an equity investor in the project. The province invested $1.5 billion directly and provided a $5 billion guarantee for project financings. More importantly, Jason Kenney, the leader of that avowedly conservative government, took the risk of Trump's re-election off the table for the other project proponents. The Alberta government would make its investment in 2020 and 2021, covering all spending incurred up to $1.5 billion. If the project was disabled, it would be Alberta — to be more specific, Alberta's taxpayers — that took most of the impact.

Kenney's administration rationalized its decision on the grounds that the project must not sit in hiatus until November 2021, and notwithstanding the potential outcome of the election of a Democratic administration implacably opposed to the project. Doubtlessly, the other project proponents put Kenney in a difficult position, demanding that Alberta preserve its schedule even in the face of that U.S. election and the risk it posed. Those other proponents could easily argue that of all the project partners, Alberta needed KXL the most, both for the increased oil sands production and for increased cash flow back into the province. Alberta stood to gain directly as the primary resource owner, and also indirectly by the improved overall economics of oil sands transportation. The other project proponents had other options, at least in theory, but Alberta needed this pipeline. If Kenney had chosen not to take the risk, not to invest, he would have rendered his province even more dependent on Canada's federal government. TMX would have remained as the only proposed large-scale pipeline project for incremental oil sands production, and Alberta would have had to bank on Trudeau's ongoing commitment to getting that pipeline built.

When Kenney decided to accept these investment terms, before the advent of Covid-19, Trump's re-election prospects were no worse than fifty-fifty. The Trump administration would, clearly, continue to do all it could to facilitate the project as long as it remained in office. Weighing the scale of long-term value to Alberta against the political risk, Kenney made a reasonable choice, and that choice indicated how much Alberta still conceived of its future prospects in terms hydrocarbon production, despite the resistance to KXL.

## LNG: Coastal Gas Link and LNG Canada

As for Alberta's other major hydrocarbon resource, natural gas, the prospect of increasing export, and therefore production, would also prove problematic, even though the regulatory processes lay entirely within Canadian jurisdiction.

In 2010 the Coastal Gas Link pipeline, owned solely by TransCanada, sought regulatory approval to move natural gas from the shale gas development in northeast British Columbia to Kitimat, on the coast. There, the pipeline would supply LNG Canada, a world-scale development led by Shell and various Asian partners who would be the product's principal buyers. Various companies proposed LNG development to transport natural gas, cooled to liquid form, from Canadian natural reserves to the coast, where it could continue to Asian markets, from Japan to Korea to China. Canadian LNG would replace some coal in those markets, thereby improving world emissions in a genuine and material fashion. Canadian production and transmission standards for LNG were as rigorous as any in the world, and minimized methane leakages, which inevitably occur to some degree during transportation. A major world-scale project on the same scale as LNG Canada was developed by Malaysian energy company Petronas, but Canadian regulatory review processes grew so protracted and problematic, specifically related to the impact of tankers on certain fish stocks, that Petronas opted to abandon the project, despite having already spent billions of dollars. By the end of 2019, sadly, of all the proposed projects, only the LNG Canada project was being constructed.

For LNG Canada's Kitimat location, Shell and its partners had successfully aligned with local stakeholders, most notably First Nations groups, to whom they provided an equity interest in the project. Also fortuitously, the project faced none of the biodiversity challenges of the kind ascribed to the Petronas project. In October 2018 the project moved forward, despite various hurdles, including fluctuations in the global LNG markets and unique national challenges of integrating Canadian gas supply into a North American gas market. The decision to move ahead with the pipeline and production facilities represented a direct investment in Canada of over US$10 billion. Moreover, by late 2018 TransCanada had achieved consensual benefit agreements with all Indigenous entities proximate to the pipeline's proposed route, including the Wet'suwet'en First Nation located just on the eastern slopes of the Coastal Mountains. These agreements included social infrastructure investments, procurement and employment opportunities, and some direct payments for de facto land access — a remarkable achievement for both parties.

The realization of LNG Canada and Coastal Gas Link created a new incremental gas market for Canadian gas producers. This was significant, because those producers had been market constrained since 2005, when the advent of American shale gas instigated a protracted period of low pricing. Coastal Gas Link would constitute a breakthrough for the industry, in some ways equivalent to what KXL represented for the oil industry. The project enjoyed the support of the Trudeau government, and also of the NDP B.C. government led by John Horgan, a marked change from that government's implacable resistance to the TMX expansion.

Global LNG supply-demand dynamics would ultimately set pricing. Canada has always had the resource base to justify world-scale LNG development, despite some concerns related to its competitiveness in periods of low pricing. But the greater risk for investors in Canadian LNG has always been the federal regulatory process and whether they could depend on that process's fundamental commitment to such development as in the national interest. The Petronas experience is a sad testimony to that reality. This despite the fact that any LNG development will serve to reduce coal in the fuel mix whether in Europe or in Asia, resulting in a lowering of GHG emissions globally regardless of which country receives the accounting credit for the reduction.

Government support of the project should have been a given since it provides both economic and environmental benefits. The scale of the investment, including the LNG production facility, transmission, and production, exceeds CDN$30 to $40 billion. A fully developed LNG development could accrue $11 billion to Canada annually until 2064. This is worth considering in the context of British Columbia's expected budgetary deficits on the order of $6 billion. The project also creates employment in Canada, as well as local taxation sources in the construction and operation phases.

But the project faced one enormous obstacle: some of the Wet'suwet'en hereditary Chiefs fiercely opposed the pipeline and refused any efforts at accommodation. Again, the elected Chiefs were aligned with Coastal Gas Link, and so was the vast majority of the Wet'suwet'en population. How governance was carried out between the elected and hereditary leaders may not have been clear to those of us observing, but those opposing the pipeline had very effective tactics. In 2013 the hereditary Chiefs and their supporters erected a fortified obstruction at an essential river crossing and allowed only minimal environmental assessment work to proceed between 2013 and 2018. Not surprisingly, this resistance allied itself with the same ENGOs that had bedevilled KXL. The environmentalists opposed any hydrocarbon development, relative emissions reductions aside; in fact, they did not accept that replacing coal with natural gas could lower emissions, arguing that potential methane leaks nullified any benefit. At the end of 2019 the construction process could not proceed without having the blockade removed, and TransCanada obtained a court order to that end. The RCMP commenced removing it, and met with various acts of disobedience across Canada, by various First Nations, in solidarity with the Wet'suwet'en hereditary Chiefs.

As a result, parts of the national rail service were disrupted, jeopardizing essential goods and services, not least of which was propane supply for basic heating across parts of northern Ontario and Quebec. The Trudeau government scrambled to find some interim fix, hoping to end the civil disobedience and avoid any unintended violence while allowing the Coastal Gas Link project to move forward, as it had the legal right to do — "legal" in the context of Canadian law.

Trudeau clearly appreciated that his government could not accept civil disobedience across the country, and that the legal rights of Coastal Gas Link and LNG Canada must be enforced. As recently as 1996, the Canadian government had negotiated a process to resolve land claims of the entirety of the Wet'suwet'en, but that did not help Trudeau's government all that much in the winter of 2020; nevertheless, on an ad hoc basis, it was able to diffuse the Coastal Gas Link situation. Protests beyond the pipeline route ceased. In mid-February, the road to the construction site was clear, and TC Energy announced that work would resume. A month later, the advent of Covid-19 would overwhelm events, but at the end of February Trudeau had achieved an uneasy stand-down.

Sadly, in February 2022, acts of civil disobedience aimed at thwarting the development of this project escalated to outright violence. At a worksite, a group of about twenty people armed with axes attacked Coastal GasLink security guards and smashed vehicle windows. The B.C. public safety minister issued a statement condemning the violence, saying the "egregious criminal activity could have led to serious injury or loss of life." The police report indicates that the workers' camp was attacked in the middle of the night by "unknown assailants wielding axes."

Trudeau chose not to make any comment, leaving it to his minister of public safety to call such actions "disturbing." All this unfolded the same week the Trudeau regime saw fit to invoke the Emergencies Act to deal with vaccine-mandate protests in Ottawa — protests that had until then remained peaceful. Were we seeing moral relativism or just politics?

## TMX

Despite a tortuous path for both TC Energy and Canada, February 2020 saw two major hydrocarbon infrastructure projects, KXL and Coastal Gas Link, under construction. That meant solace for Alberta, if not much of the rest of Canada. Again, it seemed that the realization of these projects would hinder the country's ability to meet its Paris Accord targets.

Canada's other major oil-related infrastructure project, the Trans Mountain expansion, was in its construction process as well. TMX had endured its own challenges and in 2018 was the only remaining all-Canadian export pipeline project for Alberta bitumen left standing. The project was originally led by Kinder Morgan, the world-class midstream and pipeline operator that owned the existing Trans Mountain facilities, and the scale of the expansion scale would provide over five hundred thousand barrels per day of incremental capacity, with projected capital costs of over US$7 billion.

Although Kinder Morgan had achieved full regulatory and political sanction for TMX by the end of 2016, the project continued to face various forms of obstructive litigation, ponderous regulatory enforcement of its rights to carry out the project, and jurisdictional challenges from B.C.'s NDP government. In 2017 Kinder Morgan was prepared to walk away from the project, despite having spent billions, based on its assessment that the risk of never completing the project was too great to justify further investment, regardless of how compelling the fundamental economics may be.

The stark reality of Kinder Morgan's withdrawal forced the Canadian government to acquire Trans Mountain, both its existing assets and TMX, at a cost to Canadian taxpayers of over $4 billion. The project lost an entire year of construction to accommodate a ruling from the Federal Court of Appeal, which found the first round of consultations with certain First Nations groups inadequate, especially those proximate to the Burnaby terminalling facilities. Not until mid-2019 was the Federal Court prepared to validate the project's federal approval, pending further adjustments that arose from the additional consultation process. In August 2019 construction resumed. But even with the Federal Court determination, a minority of First Nations proximate to the terminalling facilities, along with ENGOs, brought claims to the Supreme Court of Canada, claiming inadequate consultation that the regulatory and legal review processes had failed to redress. Those claims were not dismissed until mid-2020.

The Trudeau government persevered with TMX even in the face of resistance, appreciating that, if completed, the project would facilitate incremental production that could only make realizing Canada's emissions reduction targets more difficult to meet. But from an Alberta and industry perspective, the project was not lost, and was progressing comparably with KXL and Coastal Gas Link. The situation was imperfect but not dire.

The net economic benefits of TMX were comparable to those cited for KXL. TMX did not offer the same efficiency as KXL's direct route to the U.S. Gulf Coast, but it obviously did create potential for additional market diversity. By February 2022, Finance Minister Chrystia Freeland announced that the project had escalated to over $20 billion, almost 70 percent more than the earlier estimate that had predicated the government's willingness to persevere with the project. She further indicated that the project would have to look to sources of financing beyond the federal government for completion. Still, she reiterated her view that the project remained commercially viable despite the cost overrun. She did not cite reasons for the overrun, but Covid-related disruptions and the federal government's regulatory and governance burdens were the major causes. Canadian oil sands executives staunchly reiterated support for the project and confidence in its continued economic viability. With improved cost structures within the oil sands sector and global oil prices ascending to close to $100/bbl, such claims were entirely reasonable. Freeland did not elaborate on what better use those federal funds might go to — what purpose could be more valuable than completing TMX?

## Canada's Quandary

The history of these projects provides ample evidence of the intense resistance to realizing Canada's hydrocarbon potential. Opponents of the industry resorted to every available legal tactic to disrupt existing regulatory approvals, and resorted, beyond that, to outright civil disobedience. What is incredible is that the economic validity of these projects endured, as evidenced by continued creditworthy long-term commitments sufficient to finance them.

The first years of Trump's presidency provided America's hydrocarbon production industry with a federal administration utterly aligned to their interests. Of course, Trump refused to accept that human consumption of hydrocarbons caused climate change risk, and his insistent defence of that untenable position did no favours to the long-term interests of the hydrocarbon industry. Too much of the industry itself held to the denialism line, too, even as late as the end of 2019. What the United States and Canada both needed was credible and proportionate climate policy, ideally highly coordinated between the two countries. But that was inconceivable with Donald Trump in the White House. His first appointee as secretary of state, Rex Tillerson, was former CEO of ExxonMobil, but any chance Tillerson may have had to undertake reasonable climate policy was dead in the water by early 2017. With that chance died any hope that North American climate policy, under Trump, could rest on transparent, uniform, and revenue-neutral carbon pricing via a carbon tax — a real alternative to the deficiencies of the UN process and to the extremism of the environmental movement.

The Trudeau government's own climate policy ambitions depended entirely on whether the United States elected Trump for a second term. How could Canada try to impose a climate policy premised on implicit carbon pricing of $200/tonne with the United States essentially at zero? How could Canada's hydrocarbon production industry even rationalize the climate change risk without carbon pricing? And, without a continental pricing system, how could the excesses of the environmentalist left ever be countered rationally? As February 2020 rolled over into March, Canada's quandary continued: pipelines progressing even as the nation maintained its commitment to aggressive UN emission targets; meanwhile the American president denied anthropogenic climate change risk altogether. Close to 10 percent of Canadian GDP remained attributable to the hydrocarbon production sector. A defining U.S. election loomed but was soon to be overwhelmed by the advent of the global Covid-19 pandemic.

# CHAPTER 5

## Early 2020: The Advent of Covid-19

In the first days of 2020, before the advent of the Covid-19 pandemic, Donald Trump's re-election in November seemed no worse than a fifty-fifty probability. With a second Trump term the United States would not only remain outside the UN climate process but would take a fundamentally indifferent stance on the climate change risk and enable hydrocarbon production domestically while competing vigorously for export markets. Meanwhile, global hydrocarbon pricing promised to remain modest, while strong economic growth would continue across developed nations. That fundamental economic performance was the basic case for Trump's re-election, though it may not have been sufficient to overcome other elements of his record; still, a second Trump term, with its massive implications for how the developed world would deal with the climate change risk, was no worse than a fifty-fifty proposition. For Canada, the pressing question was how this country could sustain its climate position in the face of Trump's re-election.

Then, in March 2020, came the great discontinuity of my lifetime — the advent of the Covid-19 pandemic in the developed world, soon to be followed by its spread across the globe. The response to the pandemic enormously impacted global energy markets, and in turn global carbon emissions and economic output, and proved enormously instructive for the matters underpinning my argument for reconsidering decarbonization. The economic contraction we experienced throughout 2020, due to the pandemic, illustrates the kind of sacrifice required to effect even the modest emissions reductions that occurred that year. To sustain and exceed that level of emissions reductions would require massive intervention, not only to decarbonize energy systems, but to constrain basic human activity throughout the transition and likely beyond.

Furthermore, Covid-19 itself became a major factor in how the 2020 American presidential election unfolded, to eventually restore the Democratic Party to leadership under President Joe Biden. Most importantly, the world dealt with the global pandemic as an "existential crisis," thus potentially creating precedents for dealing with climate change, which was being framed as such a crisis as well.

## Markets and Covid, Early 2020

From the middle of 2015 until the beginning of 2020, pricing for most of the world's benchmark crudes settled into the range of $50/bbl to $60/bbl. This was a significant downturn from the prior fifteen years, which saw pricing in the range of $80/bbl to $90/bbl, with extended periods of over $100/bbl. Global demand had increased every year, reaching an annual average of almost 100 million barrels per day by 2019. The more moderate crude oil pricing owed not to falling demand, but to growth in global supply, most significantly from American oil shale development.

Advances in seismic and fracking technologies allowed access to a whole new supply, and also enabled a significant segment of global production growth to rest not on the vagaries of more traditional exploration processes, but on fracking process efficiencies. Momentum for this supply source grew during Obama's presidency, which — considering that his fundamental empathies lay with dealing with climate change, not with enabling hydrocarbon interests — was a testament to its compelling economic fundamentals. Then, of course, the Trump administration went on to do all it could do to maximize domestic oil production and resulting exports.

American oil shale did not compete directly with heavy oil supply, and therefore did not compete directly with Alberta dilbit, but its impact on global pricing affected all crude oils, including the Alberta oil sands, and the 2015 price adjustment in oil markets put even more pressure on the sector to gain operating efficiencies. In the previous decade, investors in the oil sands had generally operated on the accepted wisdom that, in order to generate appropriate risk-adjusted returns on investment, the price of oil must remain over $80/bbl. During that period, Alberta experienced a massive buildup of oil sands production capacity, for both mining and in situ operations. Production grew significantly from 2015 to 2019, and by the end of 2020 the oil sands were producing slightly more than three million barrels per day, even though overall crude pricing had fallen and market access was constrained. This was achieved largely by growing internal efficiencies and consolidations, along with innovations in using existing transportation outlets. These efficiencies improved the carbon intensity of the oil sands production process and in turn improved their fundamental cost structure as well as responding to emerging climate policy, and they were motivated by economics more than anticipation of some future draconian climate policy. The improved cost structure meant that even with oil prices clustered around $50/bbl, long-term capacity commitments for infrastructure such as KXL and TMX were still justified.

o o o

The story of Covid's impact on the world is an all-too-familiar one now. In the first quarter of 2020, Covid-19 began to have its effect on Europe and North America, and the International Energy Agency forecast that, as a result, 2020 would see lower global demand for oil than 2019. By how much, they could not say. Due largely to that forecast, OPEC (the Organization of Petroleum Exporting Countries) mandated production cuts to support existing but strained price levels. Russia refused to align with that mandate, and Saudi Arabia responded by reducing prices on its oil for certain buyers, down to $30/bbl — a scorched-earth retaliation against Russia's stance. By March 6, all hydrocarbon-producing countries faced price levels that strained existing production levels, let alone reasonable cash flows. Russia had failed to understand the Saudis' resolve to restore price discipline, and even if both nations resented the advent of U.S. oil shale, their failure to avoid further price erosion was counterproductive to their economic interests over the first half of 2020.

Canadian oil production was not, of course, exempt from this price collapse; in fact, the situation was particularly dire for the oil sands because of their specific transportation constraints. Benchmark crude levels of $30/bbl were not sustainable for the Alberta oil sands for any extended period. In more normal periods, low prices lead to higher demand and prices adjust upwards, but the world was about to depart radically from any normalcy. Globally, the Covid-19 pandemic would impact people's movement and consumption radically over the first half of 2020. The virus virtually dictated demand via mandated lockdowns and fears about human interaction that included conventional travel and public transport. By the end of 2020 global crude oil demand declined by about 10 percent relative to 2019, to roughly 91 million barrels. In turn, global GHG emissions declined by about 5 percent, to 34.6 billion tonnes, and global GDP declined by about 4 percent, down US$84.97 trillion. Notably, 2020 was the only year since the UN climate process began in 1992 that global emissions actually fell in absolute terms. The cost of achieving it was a mandated contraction in human activity.

The World Health Organization (WHO) had already declared a public health emergency in January 2020, shortly after the health crisis had emerged in Wuhan, China, but scientists had not yet concluded that the virus was transmitted through human-to-human contact. By the time WHO declared Covid-19 a global pandemic on March 11, significant parts of Spain and Italy were grappling with major outbreaks, and cases proliferated across all developed countries, even as governments groped to implement adequate testing and contact tracing.

On March 9, world stock markets reacted precipitously and negatively to the inescapable reality of Covid-19. The Dow Jones Industrial Average lost over two thousand points over that trading day (two thousand out of twenty-six thousand, or roughly 10 percent), and all other global exchanges incurred comparable losses. The United States, specifically, saw the end of an almost eleven-year run of rising stock market values since the 2008 financial crisis adjustment. How long would this pandemic last, and how long could any contraction of the U.S. economy be endured? Would governments act fast enough to assist large portions of the economy and electorate dislocated by this public health crisis?

President Trump addressed his nation on March 11 and utterly failed to assure markets that the United States, or the world, had any credible plan to contain, let alone end, Covid-19. But, for that matter, no other world leader of any developed economy had anything substantive to add. The grim reality set in: the world was dealing with a virus readily transmissible between humans, people had no existing natural immunity until after contracting and recovering from the illness, and a vaccine could be years away.

Most nations responded immediately with mandated isolation for all, minimizing human-to-human contact outside the home apart from essential activities — or at least what were deemed essential activities. Businesses that could go virtual quickly did. Construction, trucking, rail transport, and physical production activities had to adapt, moving outdoors when possible and mandating physical distance between workers. And still, the virus spread. By April 6 the United States reported ten thousand deaths related to Covid-19. Projected death rates validated what came to known as "stay-at-home orders." In both the United States and Canada, states and provinces, not their respective federal governments, led these mandates, primarily as a matter of jurisdiction.

Before the end of March the United States and Canada, along with Britain and the EU, passed emergency spending authorizations to ensure income for people who lost their jobs because of the pandemic, and to support impacted business and existing health infrastructure. The scale of this spending was unprecedented, and so was the speed of its implementation — virtually overnight, governments deployed trillions of dollars to finance essentially shutting down the world's economy. Coordinated action by monetary authorities across the developed world meant capital would remain available to enable this massive exercise in public debt regardless of potential inflation. To that end, the United States cut its benchmark interest rate to almost zero.

Across the developed world, Covid-19 patients overwhelmed hospitals, especially ICU units, their equipment, and, not least, their staff, and it soon became clear that people over seventy with existing medical conditions were especially vulnerable. The only way to minimize the spread of the disease was to intensify lockdowns. Elected officials depended on public health officials to take the lead in deciding how severe those lockdowns would be, relying on evolving data and imperfect modelling. No one could predict how long the pandemic would last, nor how many people it would kill, although some projections were as alarmist as 10 percent of populations if no mitigations were deployed. Mortality rates on the order of 2 percent became the overall expectation, but with great variation across countries and specific populations within countries.

The mantra "follow the science" dominated the media and most public opinion — a stance that defied Trump's unwillingness to accept the reality of the virus as much as it attested to faith in health professionals to dictate tactics. Prior to March 2020 mass "lockdowns" had not been used to combat infectious disease; empirical research had always shown that locking down a population cost too much relative to its efficacy. However, various epidemiological models had turned that opinion around radically. Most impactful was the model from Imperial College London, from a team led by Neil Ferguson. Those models predicted a "catastrophic casualty rate for an 'unmitigated' pandemic." And so, governments closed schools, cultural and sporting events, and non-essential businesses, and forced online adaptation for businesses that attempted to sustain themselves. The models were based on historic data related to influenza more so than SARS, and they tended to overestimate ultimate mortality rates incurred. The link between lockdown stringency and mortality would continue to bedevil all the subsequent Covid-19 cycles as well — not unlike climate sensitivity in the climate contest.

Some countries, mostly in the Asia-Pacific region, managed to significantly contain the virus with early enforcement of quarantine and testing and subsequent contact tracing and severe restrictions for entry. Those countries generally had more experience dealing with epidemics, going back to the 2001 SARS outbreak. North America, meanwhile, struggled to meet the demand for testing until well into the spring of 2020, in part because of design issues with the original kits distributed by the CDC (U.S. Centers for Disease Control and Prevention). This initial deficiency exacerbated the virus's spread in the United States through March and April. The CDC adjusted its test kit design, the American private sector took a bigger part in providing comparable testing, and by midsummer the United States provided testing that exceeded the numbers WHO cited as necessary for containment. The first quarter of 2020 closed out with uncertainty contributing to mass anxiety. The deaths continued to mount. By the end of April, deaths in the United States attributed to Covid-19 had surpassed eighty thousand.

While each state and province imposed its own pandemic tactics, the Canadian and American federal governments followed the same course. Neither attempted to take over authority, nor did either overtly criticize the actions of specific state or provincial authorities. The federal authorities carried out their mandates to ensure appropriate standards, approvals, and other due diligence on testing, therapeutics, and eventually vaccines. They also intervened when necessary to expedite domestic production of critical medical equipment and even emergency backstop hospital capacity. Although the Trump administration initially recommended that populations of each state remain in lockdown until it had contained the virus according to specific thresholds, some states ignored that guidance, and the federal government made no material attempt to challenge those choices.

In terms of political statements and actions, however, Trudeau and Trump differed significantly. Trudeau, like most Democratic governors and European heads of state, deferred to the health officials and other experts who held that lockdowns were inescapable in the near term, while making them as financially bearable as possible, despite whatever strain that put on the nation's finances. He stressed empathy for all and shunned any notion of premature reopening. Trump, not so much. Beginning with his first statements about Covid-19 in January 2020, he minimized the risk. Consistently, his manner and pronouncements emphasized the implications of a global pandemic for the world's economies and implicitly for his own re-election prospects. Trump took early action, banning flights from China into the United States by late January — perhaps as much out of spite as a real attempt to control the virus — whereas Canada did not take similar action for several months. Of course, the CDC's testing failures took place a month later, on Trump's watch; however, to be fair, it remains unclear how much accountability can be directed at the White House for that. Even Trump's declaration of a national emergency was couched in the expectation that lockdowns would end by Easter, less than a month away. And Trump and Trudeau were both constrained by the federal realities in play, with specific conditions related to lockdowns, and related mitigations from social distancing to masking, falling under the jurisdiction of states and provinces.

Neither Trump nor Trudeau nor any other major leader attempted to convey the potentially grim choices the pandemic might lead to. No one knew how long lockdown could continue with the global economy intact, let alone with an adequate supply of food, shelter, and electricity. No one had natural immunity to the virus, and there were no therapies for it. People could only try to avoid contact with others, and, if they were infected, to deal with the symptoms as they came. No leader, including Trump, was prepared to adopt a more Darwinian approach, for instance having people over sixty shelter in place while everyone else went about their business. In such a scenario, many people would contract the virus, some would get very sick, and a small percentage would die, but the population's natural immunity would eventually increase. Of course, such a plan would likely overwhelm hospital capacity, at least initially. In any case, grimly logical as such an approach might be, no one championed going that way, at least not directly. One might say that Trump grudgingly accepted the reality that most U.S. states would defer to their public health bureaucracies and other experts, and that each state would determine the severity of its lockdown conditions. Certain of the developed economies did adapt almost instantaneously, with many businesses and public services carrying on virtually, and others implementing social distancing. A few vital functions, such as food supply, medical services, and policing took the front-line risks in these early days of the pandemic. By the end of March, the U.S. stock market, as represented by the Dow Jones Index, had lost almost 30 percent of its peak value seen in January. This was a measure of the anxiety that Covid-19 had wrought in the developed world.

In 2020, before the advent of vaccines, resistance began growing to lockdowns and related social distancing tactics. The Barrington Declaration, which emerged in the fall, represented the most significant resistance to Covid-19 policy. The document, written by three contrarian health scientists, Sunetra Gupta of Oxford, Jay Bhattacharya of Stanford, and Martin Kulldorff of Harvard, was eventually endorsed by over 14,000 scientists, 40,000 medical practitioners, and more than 780,000 members of the public. It advocated that individuals at significantly lower risk of dying from Covid-19 — as well as those at higher risk, if they so chose — should have no constraints on their normal lives, with schools and universities open for in-person teaching and extracurricular activities; offices, restaurants, and other places of work should be reopened; and mass gatherings for cultural and athletic activities be allowed. The increased infection rate of those at lower risk would, the declaration claimed, lead to a buildup of immunity in the population, eventually protecting even those at higher risk. Those deemed especially vulnerable would be subject to some form of voluntary isolation. The declaration also emphasized the collateral damage caused by lockdown to mental health, education, and finances.

Virtually all health professionals accountable for dictating tactics relating to Covid-19 mitigation in 2020 rejected the Barrington Declaration's premises; the risk of incremental deaths, they deemed, was too extreme. Anthony Fauci, director of the U.S. National Institute of Allergy and Infectious Diseases, and lead member of the White House Coronavirus Task Force, stated:

> This idea that we have the power to protect the vulnerable is total nonsense, because history has shown that that's not the case. And if you talk to anybody who has any experience in epidemiology and infectious diseases, they will tell you that that is risky, and you'll wind up with many more infections of vulnerable people, which will lead to hospitalizations and deaths. So I think that we just got to look that square in the eye and say it's nonsense.

U.S. National Institutes of Health director Francis Collins was even more blunt and political, calling the declaration a "fringe component of epidemiology. This is not mainstream science. It's dangerous. It fits into the political views of certain parts of our confused political establishment."

Shortly after Barrington's release, the first vaccines were announced, and much of the public health debate switched, focusing now on whether efficacy justified mandating entire populations to be vaccinated regardless of any personal reservations. Notably, the public health establishment, which had closed ranks on lockdowns, was now entirely unwilling to provide any significant validation of lockdowns in cost/benefit terms of the collateral costs relative to ultimately decreased number of deaths. No political leadership across developed economies insisted on such analysis; instead, they accepted lockdowns and approved massive public stimulus rationalized to bridge the worst of the pandemic. Leaders depended on faith in the potential emergence of effective vaccines to avoid having to face the need of eventually confronting real Darwinian trade-offs between acceptable deaths and the cost of mitigating those deaths.

The parallels and distinctions between Covid-19 and climate change are elaborated in chapter 7. However, even by the early months of the pandemic a left-right schism had opened in respect of tactics to deal with Covid-19 — whether to minimize absolute deaths regardless of cost or to balance impacts on the economy. Some were willing to defer to credentialled experts even open-endedly, while others insisted that basic rights be preserved regardless of collateral impacts.

## Politics and Covid-19, 2020

One casualty of the pandemic was Trump's chance for re-election. American Democrats, along with many other "elites," from academics to media figures, had struggled from the get-go to accept that Donald Trump won the 2016 U.S. election. Trump defied what were thought to be conventional norms of opinion and behaviour for his office from the inception of his presidency in 2017 and faced persistent contempt and subversion over the course of his term; nevertheless, he held the base of his support consistently. At the end of 2019 his chances of re-election stood at no worse than fifty-fifty. His ongoing support owed as much to the state of the U.S. economy as to his actions on immigration, judicial appointments, and foreign relations. Low inflation, historically low unemployment, lower taxation, and rising average household income were realities of his record, at least up to March 2020. Nevertheless, slightly more than half of Americans viscerally despised Trump personally, regardless of his economic accomplishments, thanks to a litany of shocking actions and attitudes, such as his travel bans based on religion and ethnicity; his misguided moral-equivalence remarks after the racist Charlottesville demonstrations; his withdrawal from the UN climate process, including the commitments taken by Obama at Paris in 2015; and his border wall obsession. Even more substantively, his trade-related actions were problematic both in terms of foreign relations and basic economics.

Trump's first term approached its close with Americans polarized far beyond the traditional divide of Democrat and Republican. It would be wrong to blame Trump for that polarization; his presidency was, rather, the culmination of a fundamental division within the polity of the United States. Yet Trump's image as a champion for the populist right rested only on the most visceral of cultural issues — "social justice" matters, in particular, he simply dismissed. But many of the values Trump's presidency stood for were far from scandalous and had long roots that reached to the early 1980s: ascendant markets, enforcing the interests of private capital, central banks committed solely to containing inflation, and diminished government intervention and regulation. Though Trump took his position on climate into the untenable realm of denialism, he had the support of those who saw climate change as a risk to be dealt with, rather than as an "existential crisis" that entailed a moral imperative to decarbonize quickly and completely, no matter the cost. In sum, before March 2020, it seemed the U.S. economy might still prove the ultimate validation of Trump's governance and priorities; if he triumphed over the American centre-left and won another term, his support for the interests of the capital — notwithstanding his resort to using tariffs to resolve trade disputes, whether with China or with highly integrated partners such as Canada — would be as significant as his alignment with the populist right. At least, it might allow him to gain sufficient independent voters in key swing states.

And then came the pandemic. Significant portions of the economy shut down, with any future reopening date in the hands of health experts. For Trump's re-election prospects, this was a huge setback. The pandemic occurred on his watch, his administration failed to attempt some form of federal mandate for vigorous lockdown, his bureaucracy mishandled initial testing, and he conveyed resistance to fully delegating authority to health experts. The virus and subsequent lockdowns cratered the economy in the second and third quarters of 2020. All this was a gift to his political opposition. Trump "owned" all of it, despite that in the most populace states Democratic governors dictated the public health tactics. By the end of April 2020 almost seventy thousand Americans had died of Covid-19. In Canada, meanwhile, just over three thousand had died up to that point — much fewer on a per capita basis.

The difference in per capita mortality rates between the two countries can be attributed to a number of things — the increased use of lockdowns and other restrictions in Canada and the greater tendency of Canadians, generally, to acquiesce to government mandates regarding masking and harsh restrictions on movement and behaviour. These restrictions were necessary, governments stated, to reduce cases and, in turn, hospitalizations, and to minimize the risk of intensive care units being overwhelmed by people infected with Covid-19.

The Canadian system (which is administered provincially) is planned centrally and controls all aspects of public health care in the country, including such things as basic hospital and ICU capacity. The system ensures that no alternative private infrastructure might be provided. How the Canadian system dealt with the successive waves of the pandemic, and consequent hospitalizations, lay entirely in the hands of the health technocrats who designed the specific system in each province, admittedly, in some cases, with federal financial assistance to achieve certain minimum standards of care.

In contrast, the American system, which has evolved out of private health care, allows hospital capacity to be determined largely by the market, without intervention from even a parallel public hospital infrastructure.

Whether planned or not, the American system proved better able to deal with spikes of Covid-19 cases attributed to later variants, while Canada's hospitals, especially in Alberta, Ontario, and Quebec, came close to breaching their capacities, which in turn prompted governments to impose more closures, social distancing, and vaccine mandates.

The distinctions in these health-care systems, and their consequences during the Covid-19 pandemic, provide an illustrative example of how Canada and the United States function differently politically, with Canada relying more on public infrastructure and the United States more on private market forces. This will also impact how each country imposes climate policy.

○ ○ ○

In the 2010s, the Democratic Party's centre of gravity began shifting ever further left, away from the traditional liberalism that had typified its politicians up to that point. Not surprisingly, the Democratic Party was the only real alternative where marginalized groups could find resonance and influence; and frankly, the same was true for anyone deeply committed to immediate and drastic action on climate change. Concurrently, the Republican Party became ever less diverse, both racially and socially. The polarization of American politics became ever more fixed by the end of 2019.

By 2016, American's most marginalized had a champion in Vermont senator Bernie Sanders, a self-declared democratic socialist who launched a campaign to win the Democratic nomination for president. Hillary Clinton kept the loyalty of the party's traditional establishment majority, but it was clear just how far left the party had moved when Sanders, advocating socialized single-payer health care, wealth taxes, and radical climate intervention, attracted about 40 percent of the delegates. Despite losing to Clinton, Sanders and his base had fundamentally changed the Democratic Party. Meanwhile, Trump's 2016 mantra, "Make America Great Again," evoked powerful responses from people on both sides, and Democratic supporters deeply resented it, especially those more aligned with Sanders. The slogan clearly glorified some mythical bygone days where the political and cultural agenda was more aligned with the interests of the least marginalized of Americans. "MAGA" inherently represented, as well, resistance to the demographic change that was, in turn, changing national politics. Its adherents celebrated Trump's fundamental indifference to the marginalized.

Almost no one believed that Trump could beat Clinton in the 2016 election; the prospect seemed utterly unthinkable. Yet, in 2016, despite Clinton winning the popular vote, Trump won the Electoral College. The legacy of the Electoral College, which was created in an era that still held some skepticism about majority rule, meant that many Democratic voters were loath to accept Trump's victory; as they continually cited, more Americans had actually voted for Clinton in the popular vote, but minor shifts in the traditional voting patterns of the white working class had proved sufficient to elect Trump. There was no reason to think those same patterns couldn't prevail again in 2020 — until the pandemic. Trump's indifference and incompetence, and the mendacity in his approach, along with his failure to "follow the science" on Covid-19, mirrored his contempt for dealing seriously with climate change; he simply dismissed the evidence and advice of experts.

In terms of climate policy, the implications of a re-election were both obvious and extreme. The United States would be out of the Paris Accord formally in late November 2020, and the country would face another four years with no federal climate policy, and with more facilitation of hydrocarbon development, primarily by further reducing residual regulatory constraints viewed as unjustified and counterproductive. Over time, under Trump the U.S. federal judiciary would be transformed, and would reduce its intervention and obstruction of hydrocarbon development. Even existing federal incentives for low-carbon alternatives could have been at risk.

As the Democratic nomination process began in 2019, a double-digit array of candidates came forward, but Joe Biden was clearly the nominal frontrunner, with Sanders as his major opponent. Biden represented the Democrats' mainstream, while Sanders championed the most extreme elements of the party, who sought transformational change in America. Ultimately, the race crystallized into a contest between these two white septuagenarian men. No one could call Biden an ideologue or a policy wonk;

he was a classic professional politician, who represented and understood the particularities of the Democratic coalition, and who also proved himself flexible enough to adapt as Democratic opinion evolved. He respected the traditions and constraints of the U.S. Senate, and, while no intellectual force, was a dependable Democratic careerist in step with the party's centre of gravity wherever that moved within the political spectrum. As for climate change, Biden had shown no special interest in the issue over his days in the Senate or as vice-president, and he showed no sign of appreciating the trade-offs in play. But he had seen the demise of climate change legislation in Obama's first term up close, as vice-president, due to insufficient support in the Senate, implacable and unified Republican opposition, and a group of Democrats from states that depended economically on hydrocarbon production.

Throughout the Democratic debates held in the second half of 2019, the media treated Biden as the presumptive frontrunner. He led the polls for the nomination as of the beginning of 2020. However, at the Iowa caucuses and the New Hampshire primary in late January and early February of 2020, Biden performed badly, finishing third in each contest. It was a real question how long his campaign could continue. Then, thanks largely to his support from the African American community, Biden won the South Carolina primary in late February, and more victories followed on Super Tuesday (March 2). Virtually all the other contenders withdrew in the face of these results. Sanders alone held out, but on April 9 he capitulated and endorsed Biden, conceding that no viable path to the nomination existed for him. He said he would not run a third-party campaign for president. Sanders's acquiescence to Biden came at a price to the party and to Biden's campaign, and that became evident in 2020 and 2021, in terms of the legislative and administrative agenda that the Biden administration committed itself to, including activist climate policy.

Biden had run as the unifying restorer of normalcy and competence, never making clear how much he might, if elected, adopt Sanders's agenda. Sanders was a classic democratic socialist, with an emphasis on socialized health care, free higher education, redistribution of wealth via taxes, and reallocation of defence spending. He never was in the forefront of climate issues per se, but he clearly aligned to extreme government intervention to decarbonize the economy as soon as possible. With Biden in office, the question remained: How far would he follow Sanders's stance on climate, and how quickly? Moreover, what was that stance, exactly? Did Sanders's many supporters want and expect Biden to implement the Green New Deal, and thereby to accept decarbonizing the American economy as quickly as possible, using a massive set of market interventions to realize that objective and collaterally to reduce income disparity? And was that even radical enough to satisfy all elements of the environmental movement?

# CHAPTER 6

# The Descent to Glasgow

The Intergovernmental Panel on Climate Change issued its special report calling for 1.5°C in October 2018 and met for the Glasgow COP in November 2021. During the three years in between, the world was gripped by the aspiration of decarbonization — the same world that had agreed, at the Paris COP in 2015, to reduction strategies inadequate even to limiting a global rise in temperature to 2°C, let alone 1.5°C. North America's centre-left political leaders in particular embraced decarbonization to the point that it virtually defined their political agenda, notwithstanding both the advent of Covid-19 and the emergence of systemic racism as an issue needing urgent redress. How did the inadequate Paris targets morph into decarbonization as the world prepared for Glasgow, and what did Glasgow actually achieve?

## Canadian Credibility on Climate Targets

Canada's environmental groups fully supported decarbonization; as described in chapter 4, and in my previous books, *Dysfunction* and *Breakdown,* no Canadian hydrocarbon infrastructure project was tolerable to these groups since the advent of the UN climate process, and they could not even abide natural gas as an acceptable component of the national energy mix or export. But, of course, these groups never confronted the cost of the energy transition that they so glibly demanded — a cost that would be borne by Canadians, since energy prices would rise and there would be a loss of the revenue from activities associated with hydrocarbon extraction, processing, transportation, and sales, which would influence how Canadians could consume and produce hydrocarbons. Instead, in midsummer 2019, the government focused on passing its contentious regulatory reform agenda via Bill C-69, fundamentally changing how major infrastructure projects are reviewed and approved in Canada, providing for more expansive stakeholder participation and more ambiguous assessment criteria — initiatives that created more untenable risk for potential project developers. Not unexpectedly, no part of the federal Liberal caucus or Cabinet emerged to resist that legislation, not even those from western Canada.

The Trudeau government remained fully committed to the principles of the Paris Accord and the undertakings Canada had made in November 2015: most importantly a 30 percent reduction in GHG emissions from 2005 levels by 2030, which would require roughly a 250-megatonne reduction annually. The Canadian provinces, including Alberta, had all embraced, or at least not opposed, this as a national target. Since making that commitment, of course, Canada's emissions had in fact increased, remaining above seven hundred megatonnes annually, declining only in 2020 again due to the advent of Covid-19.[5]

From late 2015 to 2019, Catherine McKenna was Canada's environment minister. Before she was elected in the 2015 federal election, McKenna's professional experience as a lawyer had provided no expertise in climate science, the Canadian energy sectors, or even the UN climate process. Nevertheless, she was Trudeau's choice for the woeful job of publicly rationalizing the country's enduring commitment to the Paris targets, despite the fact that there was no credible case for how that might be achieved. She had to shoulder the blow to her credibility that resulted from having to serve as the champion of realizing those targets when the Trudeau government made its contentious decision to salvage the TMX expansion project, both financially and operationally, to ensure market access for expanded oil sands production over the coming decades. It didn't help her that progress on TMX and LNG Canada continued.

Admittedly, the Trudeau government fully embraced carbon pricing via a national carbon tax, but the level of that tax, $50/tonne by 2022, was not consistent with meeting the Paris targets. McKenna had no choice but to repeat the tenuous party line that Canada still had another decade to "figure out" how to achieve its reductions. In 2019, her ministry conceded that even in the best-case scenario — one combining current policies with more policies "under development but … not yet fully implemented" — Canada's total emissions in 2030 would be only 19 percent below 2005 levels. It was McKenna's special burden to continue asserting that Canada was somehow "on track." In fact, over her tenure of almost four years, carbon pricing remained Canada's only initiative that could be credited with tangibly achieving some material incremental reductions.

McKenna proved to have a problematic tenure in the environment portfolio. The environmental community never fully embraced her, either, seeing her as compromised and ineffectual. Perhaps because of the position she was put in by the actions and statements of Trudeau, her own comments were often vacuous when she was forced to detail the cost of meeting Canada's targets, and how that cost compared with what other developed nations were prepared to impose on themselves, especially in the short and medium term. She never engaged with the notion that Canada's contribution to dealing with climate change should be conditioned by the country's economic self-interest; in fact, she never engaged in genuine dialogue about the matter. It was not clear that she even understood what the social cost of carbon meant, or why it mattered to Canada's climate position. Nevertheless, she perfectly represented most left-of-centre Canadian opinion on climate and energy transition.

## Canada's 2019 Federal Election

When Prime Minister Trudeau called a federal election for October 2019, he implicitly sought electoral validation of his government's climate policies — alignment with the Paris Accord and its basic principles, coupled with carbon taxes as the pre-eminent policy instrument. The Liberals also planned to eliminate coal-based electric generation by 2030 and to initiate a national fuel standard to force low-carbon liquid fuels derived from biomass into the refinery gasoline supply mix. Although their federally mandated carbon pricing regime was not stringent enough to meet the country's Paris target, the Liberals insisted that with a gradually increasing price, plus other interventions, reductions would catch up with the goals. As for their purchase of TMX and their approval of LNG, Trudeau's Liberals offered a slightly incoherent rationalization, still invoking a vague plan to redistribute revenues from incremental hydrocarbon production to help the transition to lower-carbon energy.

In their discussions of all of these, they never really addressed how meeting the Paris target was in Canada's net economic interest. Nor did they demonstrate how the cost to be imposed on Canada compared with the costs borne by major trading partners, most notably a United States led by Trump. But then neither did the centre-left national media or academia demand that the government respond to such questions. The rationalization of the Liberal position, such as it was, was laid out in the federal document "Pan-Canadian Framework on Clean Growth and Climate Change," which was more an exhortation and statement of intent than an economic rationalization.[6] A subsequent federal analysis conceded a net negative benefit on the order of a 0.5 percent GDP annually.

In 2019 the Conservative Party, led by Andrew Scheer of Saskatchewan, advocated, in place of a climate policy, more or less unfettered hydrocarbon growth. The party rejected carbon pricing, except for large industrial emitters, and provided few specifics on how stringent that pricing would be. In fact, it supported initiatives that would increase natural gas consumption domestically, and fully supported exporting more hydrocarbons, especially LNG. Scheer still, however, remained nominally committed to Canada's Paris targets. The Conservatives' credibility on climate was even more compromised than the Liberals'.

Trudeau won a minority government despite the fact that the Liberals' share of the popular vote was actually about 1.5 percent lower than the

Conservatives'. The defeat of the Conservatives was attributed by a number of commentators to the fact that they lost 2 percent to the more extreme right-wing People's Party. But that was cold comfort for Canadian conservatives. Not surprisingly, the popular vote revealed an extreme regional polarization between most of urban Canada and regions that depended on hydrocarbon production as a key economic driver.

Admittedly, the Conservative position on climate policy was not tenable, incapable as the party was to even accept carbon pricing as foundational to credibility. However, the Liberal position was far from optimal, let alone reasonable, either. The striking fact was that during the election campaign, neither party was challenged to justify its position in terms of cost. The Canadian body politic was not, apparently, interested in intellectual honesty in the arena of climate policy, and nurtured a continuing disconnect or just plain lack of interest in the economic contribution of hydrocarbons to the Canadian economy and the net cost to Canada of a transition from hydrocarbons. But at least as 2019 ended, the Liberals never explicitly committed to net-zero emissions. They merely adhered to their Paris national emissions reduction target, which those attentive to the UN process recognized as insufficient, along with the targets of the other developed nations, to even approach the avowed aspiration of 2°C containment.

## Failure in Madrid

In late 2019, COP's twenty-fifth meeting took place in Madrid, to finalize the rules on how countries could reduce their emissions using international carbon markets, as laid out conceptually under Article 6 of the Paris Accord. Carbon markets in this context were premised on the notion that some countries would achieve more emissions reductions than the Paris Agreement required of them and could therefore sell these excess reductions to other countries deficient in meeting their obligations.

Article 6 proved extremely difficult to implement because it was not clear what should count as a legitimate incremental emissions reduction, and consequently, what country should receive credit for such an emissions reduction. But the concept of incenting emissions reductions globally on the lowest cost basis remained appealing, regardless of where such emissions came from. If the cost of buying an emissions reduction from a less developed country was less than the cost of generating it in a more developed country that had an obligation to reduce emissions, then a real economy was created globally.

A classic example of this is found in LNG trade, where natural gas substitutes for existing coal utilization. Consider Canadian LNG trade to China. Canada sells LNG to meet Chinese power demand, which displaces coal consumption in China and reduces emissions in China. That reduction is only possible if China imports the LNG from Canada, but Canada itself increases its national emissions producing the LNG. The world gains from an emission perspective, but should China get all the credit and Canada get none? The planet wins, but Canada's emissions position worsens. The same logic applies to any country that might become the LNG supplier to China.

Understandably, Article 6 alienated Canada's hydrocarbon sector. Since the UN process's inception, carbon emissions had been allocated to the country where they occurred. This may seem intuitively obvious: Canada combusts oil and gas, so it owns the resulting emissions, both from producing the oil and gas and from transporting it. But LNG export to Asia had real economic value for Canada, as validated by the various projects developed over the previous ten years, including LNG Canada, which was proceeding to operations. Canada would incur incremental emissions from these projects, mostly through methane emissions during natural gas production and transmission processes, and in generating the energy to compress that gas to liquid. Emissions reductions would occur in Asia, as coal was reduced in the electric generation fuel mix. Logically, then, the two countries involved should share the value of the reduction. Unfortunately, the Trudeau government, via its new environment minister, Jonathan Wilkinson of North Vancouver, had difficulty advancing that position at Madrid. And why? Wilkinson proffered the absurd contention that sharing reductions with Asian countries in this way would diminish the effort for physical emissions reductions within Canada. That was neither demonstrably true nor even relevant to the real issue of who gets credit for genuine emissions reductions.

Unsurprisingly, Wilkinson's failure to champion the notion of, at a minimum, sharing any global emissions reduction arising from LNG for coal substitution between countries further alienated the country's hydrocarbon interests from the Trudeau regime. Framing Article 6 strictly in terms of a potential source of emissions compliance credits for Canada hardly advances the country's economic interests, but it is predictably consistent with a bias against Canadian hydrocarbon development, even when that development reduces global emissions.

This fundamental issue of who gets credit for an actual emission reduction and whether the reduction can actually be sold as an "off-set" remains problematic to the entire UN process. The LNG example, although so pertinent to Canada, is not the only one. Deforestation is de facto an emission, as it reduces natural sinks for emitted carbon, but it rarely, if ever, is treated in the same terms as combustion. Intellectual rigour has never really been applied. Other considerations more related to redressing inequality and historic culpability distort that objective.

The Madrid meeting extended to over two weeks, but even with the extra time, the parties failed to set accounting rules to govern the Paris Agreement, stifling progress on a global carbon market. UN secretary general António Guterres said he was "disappointed" with the result of COP25 and that "the international community had lost an important opportunity to show increased ambition on mitigation, adaptation, and finance to tackle to climate crisis." Indeed, the meeting achieved no increase in voluntary emissions, and it was decided that adoption of Article 6 would have to wait for further discussions in Glasgow, scheduled in late 2020. Once again, the UN climate process exhibited the inherent difficulty in achieving alignment across more than 190 countries, each with its specific agenda and gamesmanship.

None of this deterred Environment Minister Wilkinson from announcing late in 2019 that Canada would introduce legislation to enshrine the goal of net-zero emissions by 2050, stating that "we will no longer be using fossil fuels." However, he announced no immediate adjustment to Canada's target.

## The Green New Deal

During 2009 and 2010, the first two years of the first Obama administration, the Democrats held majorities in both House and Senate, even, for a time, having sixty senators, which made them theoretically filibuster-proof provided Democratic solidarity prevailed. During that time, the House Democrats made a serious attempt to advance comprehensive federal climate legislation, with key provisions to put a price on carbon using a national cap-and-trade system. In 2009 the House of Representatives even passed a bill known unofficially as Waxman-Markey (after the two congressmen who sponsored the bill), albeit by the narrow margin of 219 to 210. Formally, it is known as the American Clean Energy and Security Act. The bill mandated emission caps that would reduce aggregate GHG emissions to 17 percent below 2005 levels by 2050 — a goal consistent with the Obama administration's intentions on reduction targets to be submitted at the 2009 Copenhagen Climate Conference.

The bill mandated a cap-and-trade scheme, and Congress found it difficult to agree on how emissions allowances were to be allocated; however, a difficult compromise, one relating mostly to the generous treatment of coal interests, was reached, and the bill was passed. It never became clear what carbon price would have emerged had it been enacted, since the bill died in the Senate. It needed the votes of sixty senators, and they never materialized; nor did sixty senators ever support any alternative cap-and-trade formulation. Solid Republican resistance, coupled with Democratic resistance from states still levered to hydrocarbon production, proved impossible to overcome. It didn't help that the Obama administration throughout 2009 and 2010 prioritized health care and financial reform over climate policy.

The environmental movement saw an enduring lesson in Waxman-Markey's near success and ultimate failure. From their perspective, this was a simple, market-based bill that would have required polluters, rather than citizens, to pay for the transition from fossil fuels to renewable forms of energy; but the legislative process had made it impossible. A straightforward bill with a purpose had been doomed by corporate interests obsessed with finding competitive advantage in the allocation process. All this seemed to show a fatal flaw in the very notion of market-based climate bills, especially if they were of cap-and-trade design.

Beginning in 2010, and for the next eight years, the Republican Party controlled both the House and the Senate, so no further federal climate legislation came forward. In 2017, the Trump administration considered employing a national carbon tax to facilitate other tax cuts, but, sadly, nothing came of those deliberations. Of course, in October 2018, IPCC published its special report on 1.5°C containment, and shortly thereafter in November 2018 the Democrats regained control of the House of Representative by a substantial margin.

In February 2019, Representative Alexandria Ocasio-Cortez of New York and Senator Edward Markey of Massachusetts began promoting what they termed the Green New Deal (GND), in the form of a Congressional resolution. The name, of course, directly referenced Franklin Delano Roosevelt's massive economic interventions during the Great Depression, when his New Deal created most of the social safety net that persists in the United States to this day. Fundamental to GND's impetus was the notion that the climate crisis required Democratic intervention and radicalism equivalent to what the Great Depression once demanded — the climate crisis, that is, as defined by the 2018 IPCC special report on 1.5°C containment. Ocasio-Cortez and Markey became the public faces of the GND, but it also had the support of the more left-leaning elements in the Democratic caucus, both House and Senate, amounting to roughly one hundred co-sponsors.

Those co-sponsors joined forces with environmental activists to develop the specific terms of the GND. Its core intention was to decarbonize the American economy as fast as possible. Since the defeat of the American Clean Energy and Security Act, no Democratic legislative plan had been proposed to deal with climate change, and in the interim, few Democratic politicos have endorsed decarbonization as explicitly as the GND now did. It called for 100 percent clean, renewable energy from zero-emission energy sources nationwide by 2030, and for electric generation and net-zero emissions across the economy by 2050. It specifically cited other complementary initiatives, such as digitizing the nation's power grid, upgrading buildings for higher energy efficiency, and investing in electric vehicles and high-speed rail. It even referred to mitigating emissions from the agricultural sector, especially from cows.

The sponsors rationalized its interventions by citing the cost of potential extreme weather events, as estimated by the fourth National Climate Assessment Report, authored by the U.S. Global Change Research Program (USGCRP) under the oversight of the National Oceanic and Atmospheric Administration (NOAA). That figure was $500 billion in economic loss in the United States alone, each year, by 2090. Notably, the report included no estimate of the social cost of carbon to contextualize the damages estimate.

Predictably, virtually any public figure who could be identified as left of centre within the American political spectrum and cared to comment on the Green New Deal proved rhapsodic. Rhapsodic and unthinking. Naomi Klein's comments in *On Fire: The Case for the Green New Deal* are entirely representative:

> What we cannot do, under any circumstances, is precisely what the fossil fuel industry is determined to do and what your government is so intent on helping them do: dig new coal mines, open new fracking fields, and sink new offshore drilling rigs. All that needs to stay in the ground.
>
> What we must do instead is clear: carefully wind down existing fossil fuel projects, at the same time as we rapidly ramp up renewables until we get global emissions down to zero globally by midcentury. The good news is that we can do it with existing technologies. The good news is that we can create millions of well-paying jobs around the world in the shift to a post-carbon economy — in renewables, in public transit, in efficiency, in retrofits, in cleaning up polluted land and water.

In their article, "Green New Deal: Ocasio-Cortez Unveils Bold Plan to Fight Climate Change," published in the *Guardian* on February 7, 2019, Emily Holden and Lauren Gambino astutely observed how the Democratic Party's position on climate had moved to the left since Waxman-Markey:

> The resolution says it is the duty of the federal government to craft a Green New Deal "to achieve net-zero greenhouse gas emissions." That includes getting all power from "clean, renewable and zero-emission energy sources."... The document also endorses universal healthcare, a jobs guarantee and free higher education — a huge shift in messaging from nearly a decade ago, when Democrats were advocating for a cap-and-trade system to limit greenhouse gases by allotting industry permits for pollution.

There was negative reaction to the GND. Greenpeace co-founder Patrick Moore perfectly captured the contempt for the glib extremism of the Green New Deal, attacking Ocasio-Cortez: "Pompous little twit. You don't have a plan to grow food for eight billion people without fossil fuels, or get the food into the cities. Horses? If fossil fuels were banned, every tree in the world would be cut down for fuel for cooking and heating. You would bring about mass death."

In March 2019 the U.S. Senate, controlled by the Republicans, voted on the GND resolution. It rejected it with a vote of fifty-seven to zero. Clearly, even the Senate Democrats were not ready to embrace the GND's fundamental tenets: forty-three Democrats voted "present," the equivalent of abstaining.

Still, a substantial portion of the Democratic House caucus had laid down its marker on climate change, and that would impact the presidential nomination process within the party. Prior to this vote, Speaker Nancy Pelosi had been equivocal; she never advanced the GND in any serious legislative fashion, no doubt because she knew there was not enough support even for the House of Representatives to advance it. Still, a significant component of the Democratic base was fully aligned with the GND's fundamental ambitions, and it remained a standard against which all contenders for the Democratic presidential nomination would need to react. That process, of course, came down to a choice between Bernie Sanders and Joe Biden — a choice between unapologetic democratic socialism and classic brokered

Democratic liberalism, influenced crucially by weighing who had the best chance of defeating Donald Trump.

Sanders had no issues with authenticity; his position had remained clear and consistent throughout his career as a leftist agitator and politician. By midsummer 2019 Sanders had taken a climate position equivalent to the Green New Deal resolution, that the United States must decarbonize transportation and power generation, its two largest sources of emissions, by 2030, thus lowering emissions by 71 percent. He implied that the government would take over or constrain the existing electric generation and transportation sectors, with associated federal spending in the range of $16 trillion. Castigating the hydrocarbon industry, Sanders stated that it was the moral equivalent of the tobacco industry. He intended to lead litigation, he said, against the industry for its historical culpability in resisting action on climate change and for collateral impacts.

Biden, conversely, was a career Democratic politician, committing himself more to following the party's centre of gravity than to any specific ideological agenda. In the heat of the first presidential debate with Trump, in October 2020, he stated, "I don't support the Green New Deal." But what did he advocate, and how different was it from the GND? His key promise was to ensure that the United States achieved a 100 percent clean energy economy with net-zero emissions by 2050, using an unspecified mix of enforcement, direct federal investment, and fiscal incentives. He fully endorsed the conclusions of the IPCC special report on 1.5°C and committed to eliminating carbon emissions from the electric sector by 2035, imposing stricter gas mileage standards, funding investments to weatherize millions of homes and commercial buildings, and upgrading the nation's transportation system. To reach its goal of carbon-free electricity by 2035, Biden's plan included wind, solar, and several forms of energy not always counted in state renewable portfolio standards, such as nuclear, hydro power, and biomass. All this amounted to spending on the order of $2 trillion.

Overall, his climate policy served its required political purpose; it avoided alienating the Democratic Party's more extreme element, and sufficiently assuaged the more moderate elements that Biden's administration would restore the country's commitment to dealing meaningfully with climate change — in stark contrast to the Trump administration's climate change denialism. If not the GND, per se, Biden put forward the most ambitious, if not extreme, climate position of any Democratic presidential nominee in history.

## Credibility: Canada and $170/Tonne

Joe Biden was inaugurated as president of the United States on January 24, 2021. To say this was a relief to the Trudeau regime is doubtless an understatement. A Trump re-election would, among many other challenges, have left Canada committed to a UN process rejected by the United States, at least in terms of the Paris Accord, thus intensifying Canada's quandary as it tried to approach its targets without further straining bilateral competitiveness vis-à-vis the United States.

Anticipating Biden's inauguration, the Trudeau government had, in December 2020, announced major adjustments to its climate policy. Specifically, the government's carbon price would increase from $50/tonne in 2022 to $170/tonne by 2030. The government also pledged $15 billion of new spending relating to climate over the coming ten years, including $3.2 billion for reforestation, $1.5 billion for developing low-carbon fuels, and $3.5 billion on efforts "to rapidly expedite decarbonization projects with large emitters." However, the announcement included no specific mandates on how Canadian would be allowed to use energy, and no specific constraints on hydrocarbon production beyond those implicit in the existing oil sands emissions cap put in place by Trudeau and Alberta premier Rachel Notley in 2015.

The December 2020 announcement made no reference to the economic value of those initiatives, even in the context of a $170/tonne carbon tax — a real deficiency. Nevertheless, Wilkinson got to make the announcement that Catherine McKenna had doubtless yearned to make — Canada had a plan to materially increase its credibility on meeting its Paris emissions reduction targets. A $170/tonne carbon tax would make a significant impact, if Canada ultimately had the will to impose it on itself. The government's own modelling emphasized, however, that most Canadian emissions reduction would occur in the hydrocarbon sector, from an unspecified combination of demand-destruction for its products, diminished production due to higher costs, and CCS.

The announcement did not clarify how Canada would remain competitive, if indeed it could, at this level of carbon pricing. Wilkinson contended that the government would explore a more robust system of "border adjustments" to ensure that Canada retained its market share of energy-intensive exports and that foreign imports would not gain an advantage over Canadian products that had to internalize the higher carbon pricing. The country would not be hindered by a more stringent climate policy than those of the countries it trades with — that was the promise. Border adjustments would be represented by tariffs imposed on foreign imports to account for the difference in carbon pricing, and conversely, Canadian entities would receive rebates to make up for carbon taxes paid on exports into countries that did not have comparable carbon pricing.

The issue of competitiveness promised to become more significant and problematic as Canada's carbon tax increased relative to the prevailing U.S. carbon price. If the United States did not have a transparent federal carbon tax, akin to that in Canada, even determining the prevailing U.S. carbon price would be problematic.

The additional cost imposed by this $170/tonne carbon tax level and related national initiatives to meet Canada's Paris emissions reduction obligations should have been contextualized and rationalized in the context of the net benefit that accrues to a country as part of a coordinated global effort to reduce climate change risk. Wilkinson provided no direct comparison of higher energy costs relative to projected insurance claims from prospective extreme weather events, leaving unclear whether he recognized the real problem of free riders such as China, India, or even the United States. Canada's commitment to $170/tonne, as articulated, was unconditioned, with no off-ramp if, for instance, those countries make no comparable carbon pricing or equivalent climate policy commitment. To make matters worse, a still higher carbon tax may be necessary to achieve Canada's Paris emissions reduction targets.

Western Canadian governments, especially Alberta's United Conservative Party government, had long contested the federal government's right to impose a federal carbon tax, on the basis that environmental issues fell under provincial jurisdiction. On its face, this argument was never compelling; climate change was clearly a global issue, and by extension a national one. Still, Alberta resisted the federal carbon tax, seeking resolution from the Supreme Court of Canada. On March 25, 2021, the Court released its much-anticipated decision, upholding the constitutionality of the Greenhouse Gas Pollution Pricing Act, the centrepiece of the federal government's climate change plan, which imposes minimum carbon-pricing standards on the provinces. The majority of judges in the six-three split decision emphasized the importance of a national approach to addressing climate change. A $170/tonne price would be applied regardless of Alberta's objections, and any change to that would have to be decided politically, not by the courts. This decision affirmed the constitutionality of federal climate policy, an important step forward. The Trudeau government, then, achieved additional credibility on climate policy generally and on meeting Canada's Paris targets.

## Carbon Taxes in the United States

By the end of 2019, it was clear that Canada planned to adopt carbon pricing, via carbon taxes, as its pre-eminent climate policy instrument, with a current and ultimate stringency that few other jurisdictions had yet to achieve or even seemingly aspire to, despite California and the EU having had cap-and-trade systems in place over much of the previous twenty years. Ideally, Canada's carbon pricing would be consistent with the policies of its major trading partners, especially the United States. But even with Biden as president and significant Democratic control of the U.S. Congress, carbon pricing has not been incorporated as a major policy instrument for climate goals.

Despite inaction federally, certain states did implement initiatives. Throughout the Trump presidency, California positioned itself as the point of resistance to his agenda, and particularly in respect of climate. In 2018 the California state legislature enacted SB100, thereby committing itself to

100 percent carbon-free electricity by 2045, 50 percent renewables by 2025, and 60 percent renewables by 2030. The legislation amounts to a mandate imposed on its electric generation utilities.

By the end of 2019, California had already approached almost 30 percent of its installed electric generation capacity in the form of renewables. The 2045 goal was framed expansively as "carbon free," meaning that California could restore nuclear and natural gas combined with CCS systems, despite resistance from certain environmental activists within the state.

California's cap-and-trade system had already been in place for fifteen years by 2019, but it had never generated carbon pricing much above US$20/tonne, even for its large industrial emitters. Pricing was not California's pre-eminent policy instrument; in fact, it had little effect besides providing revenue to the state. California's choice of cap-and-trade, rather than a carbon tax, was originally premised on the greater certainty it provided of a specific level of emissions reduction rather than certainty on the price to emit. It also functioned as a complement to other regulations (fuel economy standards, building codes, renewable energy mandates) that effected emissions reductions.

That was the theory, but, in fact, the cap-and-trade system produced only about 15 percent of the state's emissions reductions up to 2020. The rest came from sector-specific mandates akin to the one applied to electric generation in SB100. The California cap-and-trade system was complex and administratively burdensome for the major industrial emitters that had to comply with it, and it was not very efficient. Yet, even into 2021, with Democratic control of both the governor's office and state legislature, there was no significant support for a transparent carbon tax for the state.

In the state of Washington, one of the country's most liberal, voters rejected the notion of a state-wide carbon tax applicable to almost all state emissions in a 2018 referendum by a margin of 56 to 44 percent. Incredibly, much of the state's environmental community opposed the proposal. The tax would have started at US$15/ton in 2020 and then increased by $2/ton per year (plus inflation) until 2035, when it would reach, depending on inflation, around $55/ton. Even its advocates recognized that the pricing strategy was inadequate to drive transformational change in energy utilization but backed it on the grounds that the state would redeploy the tax revenues on low-carbon initiatives and to reduce specific climate impacts.

Two years before, a proposal for a carbon tax designed to be more stringent while also being revenue neutral for the state government had failed, because the more radical elements of the state's environmental movement had resisted the concept of revenue neutrality and fundamentally preferred regulation and direct intervention to the more indirect economic processes implicit in carbon pricing. Washington's 2018 carbon tax proposal was designed by more pragmatic elements of the state's environmental groups and dropped tax revenue neutrality. The pricing was more moderate, and the carbon tax collection would fund more low-carbon investment. The hope was that a less purist form of carbon tax would be more favourably accepted by the Washington state electorate. When it was rejected, its advocates faced the chilling reality that a majority of the electorate, when faced with paying directly for climate change mitigation, opted for minimizing impacts on their household expenses. The transparency and intellectual honesty of the carbon tax became its own political liability.

Nationally, the notion of a federal carbon tax received some attention from Trump administration officials who were trying to offset the proposed Trump tax cuts of 2017 with some other revenue source. But despite the significant revenues that could have been generated even at pricing levels close to those imposed in Canada at that time, the concept was never seriously acted upon. Even after the Democrats took control of the House of Representatives in late 2018, no climate legislation based on carbon pricing came forward. The GND did not rely on carbon pricing, sadly.

For context, at this time the IMF was calling on all developed countries to implement an average global carbon price of US$75/ton by 2030. This had no impact on U.S. politics vis-à-vis carbon pricing.

## Elites, Governance, and Climate Change

The World Economic Forum, headquartered in Geneva, Switzerland, holds an annual meeting in Davos, typically each January, for the world's political and business elites to discuss major issues and risks that impact the global economy. The Davos agenda essentially reflects a consensus of what comprises the key risks. And by 2020, the issue of climate change dominated those deliberations. All five of the risks deemed most urgent related to climate, including extreme weather events, failure to slow emissions growth, insufficient investment in adaptation, and biodiversity loss in the most vulnerable of nations. Emphasis was placed on the potential damages caused by extreme weather events, which could become uninsurable and over time lead to stranded assets disrupting existing global pension arrangements and general financial stability. All businesses were exhorted to internalize the climate change risk into their planning explicitly, as part of their basic internal governance. The Davos proceedings showed some despairing recognition that global energy demand continues to grow, as certain developing nations are not yet prepared to constrain their national economic development and, in turn, their hydrocarbon consumption.

The World Economic Forum, led by Canadian Mark Carney, Goldman Sachs alumnus and former central banker for both Canada and Britain, became the central advocate for the "Great Reset." It was proposed that in a post-Covid-19 world, governments would have to modify capitalism to serve "stakeholders," as opposed to simply delivering financial results to "shareholders," who invested capital. This would apply especially to climate. How the world dealt with Covid-19 would, the attendees argued, provide a template for dealing with climate change: defer to scientific experts, emphasize higher levels of risk mitigation, and reduce economic inequality. They pointedly did not emphasize economic growth and efficiency. Growth would need to be greener and fairer. Biden and the Congressional Democrats rebranded the Great Reset as "Build Back Better," showing their solidarity with the basic notions of "green" and "fair" as their guiding ethos. In so doing, they rejected the guiding principles of deferring to markets and reducing regulatory constraints, which had animated the long period of sustained global growth since the early 1980s.

Carney was appointed as United Nations special envoy for climate action and finance in March 2020. In that role, his greatest accomplishment to date has been the creation of the Glasgow Financial Alliance for Net Zero, which has a basic objective of generating US$100 trillion through 2050 to realize investments to effect lower emissions while still earning a profit. His credibility has been questioned by some since he has a continuing role as vice-chairman of Brookfield Asset Management, a Canadian investment firm that is a long-time investor in fossil fuel infrastructure. Despite that "inconvenient truth," he maintains great influence as a philosophical rationalizer of decarbonization and a fierce critic of existing capitalist economies, most notably through his public commentary and writing.

In his book *Value(s): Building a Better World for All*, Carney writes, "We are approaching the extremes of commodification, as commerce expands deep into the personal and civic realms. The price of everything is becoming the value of everything." Climate change he calls "the Tragedy of the Horizon," explaining, "we don't need an army of actuaries to tell us that the catastrophic impacts of climate change will be felt beyond the traditional horizons of most actors — imposing a cost on future generations that the current generation has no direct incentive to fix." What we need, therefore, he contends, is "clear, credible and predictable regulation from government. Air quality rules, building codes, that type of strong regulation is needed. You can have strong regulation for the future, then the financial market will start investing today, for that future. Because that's what markets do, they always look forward." He does admit, however, that this transformation "wouldn't happen spontaneously by the financial sector."

Regardless of the merits of his critique on the fundamental utility of free market capitalism, Carney rightly recognizes that governments, especially those with democratic mandates, have both the legitimacy and capacity to set the terms around which markets should or should not internalize the risk of climate change. He insists, also, that they have the obligation. But in his perfect world, the realm of corporate governance faces interventions that attempt to force the internalizing of the risk of climate change, if not the inevitability of decarbonization, regardless of where specific governments stand in respect of climate policy.

The most famous example of the kind of corporate action Carney advocates so far comprises the actions of certain key financial asset managers, most notably investment management company BlackRock, which is one of the top three shareholders of more than 80 percent of the companies in the S&P 500. This position enables it to vote on shareholder resolutions directly, without consulting the various individuals with financial interests in their many funds and other products. The company's imperious CEO, Larry Fink, is a major Democratic donor, who in mid-2020 wrote an open letter to American CEOs, conveying that BlackRock would be "increasingly disposed" to vote against companies that fail to report climate change impacts and/or demonstrate clear plans to address them — presumably in a manner acceptable to BlackRock. Such votes could take various forms, from specific shareholder resolutions on company strategic policy, nominations for board seats, or support for existing management teams, or even the ultimate "vote" of removing the investment position entirely. These demands often extend beyond how firms currently internalize the risk in their strategic deliberations, such as assuming accountability for the impact of their products on third parties, clients, and others, all the way down the supply chain. This BlackRock initiative shows how the ESG (environmental, social, and governance) trend obscures who holds the accountability for corporate strategy, leaving corporate management struggling to find an appropriate balance between meeting expectations on the internalizing climate risk and preserving value.

The audacity of the ESG is perfectly captured in the various letters to CEOs that Fink has issued in recent years. Some classic examples:

> The world is moving to net zero, and BlackRock believes that our clients are best served by being at the forefront of that transition. We are carbon neutral today in our own operations and are committed to supporting the goal of net zero greenhouse gas emissions by 2050 or sooner. No company can easily plan over thirty years, but we believe all companies — including BlackRock — must begin to address the transition to net zero today.
>
> It's important to recognize that net zero demands a transformation of the entire economy. Scientists agree that in order to meet the Paris Agreement goal of containing

> global warming to "well below 2 degrees above pre-industrial averages" by 2100, human-produced emissions need to decline by 8–10% annually between 2020 and 2050 and achieve "net zero" by mid-century.
>
> Given how central the energy transition will be to every company's growth prospects, we are asking companies to disclose a plan for how their business model will be compatible with a net zero economy — that is, one where global warming is limited to well below 2°C, consistent with a global aspiration of net zero greenhouse gas emissions by 2050. We are asking you to disclose how this plan is incorporated into your long-term strategy and reviewed by your board of directors.
>
> And BlackRock does not pursue divestment from oil and gas companies as a policy. We do have some clients who choose to divest their assets while other clients reject that approach. Foresighted companies across a wide range of carbon intensive sectors are transforming their businesses, and their actions are a critical part of decarbonization. We believe the companies leading the transition present a vital investment opportunity for our clients and driving capital towards these phoenixes will be essential to achieving a net zero world.

It is important to note that in neither Fink's 2020 nor his 2021 CEO letter does he acknowledge that BlackRock's commitment to decarbonization requires the virtual elimination of the hydrocarbon production sector. Or that under-investment in hydrocarbon production and related infrastructure over previous years, in part due to ESG decarbonization advocacy, contributed to the energy crisis in Europe over the winter of 2021 and 2022, as well exacerbating geopolitical tensions in the region. He certainly not does discuss the net cost or benefit of decarbonization as an issue deserving of continuing debate, nor balance in how corporations attempt to project the ultimate policy response.

In May 2021, three of the world's major integrated hydrocarbon production entities were impacted by this phenomenon. First, ExxonMobil lost a proxy battle after certain shareholder activists, supported by some key asset managers, including BlackRock, found the company's approach to energy transition deficient. As a result, ExxonMobil saw three dissident directors appointed to its board. That same week, the International Energy Agency (IEA) warned that investment in new fossil fuel projects must stop immediately, to allow the global energy sector to achieve carbon neutrality by 2050. ExxonMobil did not embrace that directive as its basic corporate strategy before or after the shareholder vote, but the vote's outcome was still a measure of how certain business elites and climate activists had collaborated to press the imperative to transition basic energy systems, seemingly without consideration for shareholder value or what constitutes optimal climate and energy policy.

On the same day as ExxonMobil's vote, Chevron's shareholders passed a resolution to reduce emissions from the end use of its products. Emissions from gasoline, jet fuel, and petrochemical consumption cannot be reduced materially over the short run, and perhaps not at all in some cases, other than by reducing demand. The resolution was patently absurd without a fundamental reinvention of Chevron's fundamental value proposition, but nevertheless, it gained 61 percent shareholder support. Finally, contemporaneously, Royal Dutch Shell was ordered by a Dutch district court to amend its corporate strategy to slash its emissions harder and faster. Not surprisingly, Shell responded by appealing and relocating its head office to the United Kingdom.

For decades, these entities, along with BP, had dominated the world's private energy production and processing sectors, funding fundamental research into the nature of climate change risk and possibilities of transitioning to less emissive fuels. The assaults on their basic raison d'être as corporate entities are emblematic of the radical expectations implicit once the goal of decarbonization is endorsed as some moral imperative to mitigate climate change, regardless of the impacts on shareholder value, not to mention global human welfare. But for the self-avowed elites to so glibly embrace decarbonization without acknowledging its net cost or benefit, and adjusting public policy accordingly, can only remind one of George Orwell's words, "One has to belong to the intelligentsia to believe things like that. No ordinary man could be such a fool."

## IPCC Assessment Report 6

In August 2021, the IPCC released its working group report on physical science as part of its Sixth Assessment Report on climate change. Running to four thousand pages, it was the first such report of AR6; the others, covering impacts, adaptation, and mitigation, will appear in 2022, along with the final synthesis report. That first report came out only months before the Glasgow COP26 meeting, and essentially reiterates the admonishments of the 2018 IPCC special report on 1.5°C temperature containment.

Here I recount some of its key findings for emphasis. Global mean surface temperature has increased by 1.09°C since the pre-industrial baseline of 1850 to 1900, and the report projects that global temperature will be 1.4 to 4.4°C hotter than in pre-industrial times by the end of this century, depending on whether emissions are rapidly cut to net zero or continue to rise. The 1.5 and 2°C limits cited in the Paris Agreement will be breached, the report goes on, unless $CO_2$ and GHG emissions are rapidly cut within the next five to fifteen years. However, it concedes that climate sensitivity, that is, temperature increase as a function of additional GHG concentration, remains within a significant range — though narrower than what was previously considered probable — so that it is still uncertain how extreme the warming effect will be. Not surprisingly, the report concludes that human influence is responsible for the climate change since 1850.

On the matter of "tipping points," the report is more equivocal. It defines a tipping point as "a critical threshold beyond which a system reorganizes, often abruptly and/or irreversibly." Our global climate system is currently responding to warming "proportionate to the rate of recent temperature change," the report says, but "some aspects may respond disproportionately." It makes the important statement, however, that there is no evidence that such a tipping point may occur in the next hundred years. About extreme weather events, such as floods, storms, and forest fires, along with extremely hot and extremely cold spells, the report adds that "regional changes in the intensity and frequency of climate generally scale with global warming." The only way forward, it prescribes, it to decarbonize globally: "Without net-zero $CO_2$ emissions, and a decrease in the net non-$CO_2$ forcing (or sufficient net negative $CO_2$ emissions to offset any further warming from net non-$CO_2$ forcing), the climate system will continue to warm."

As for net negative emissions, the report cites "the potential to remove $CO_2$ from the atmosphere and durably store it in reservoirs," which could be "implemented at a large scale to generate global net-negative $CO_2$ emissions," resulting in "anthropogenic $CO_2$ removals exceeding anthropogenic emissions" that would "compensate for earlier emissions as a way to meet long-term climate stabilization goals after a temperature overshoot." This notion is heartening but not emphasized.

## The Biden Presidency and Climate Change

Biden was elected president of the United States on November 3, 2020. In the first hours of his terms, Biden terminated the KXL pipeline by revoking its presidential permit, and he closed off leases in federal lands a few weeks later; by executive action, he did as much as possible to signal his climate bona fides and in turn impair the hydrocarbon sector production and growth. He could do no more without federal legislation.

Following the election, the Democrats also obtained nominal control of the U.S. Senate and could advance their agenda on both climate and expanding the American welfare state — albeit only with the vote of Vice-President Kamala Harris, to create a majority of fifty-one to fifty. The Democrats had control of House of Representative, the Senate, and the White House — potentially a historic opportunity to realize an agenda that synthesized Biden's campaign platform with many of the core elements of Sanders's social policy and the GND.

Of course, the U.S. Senate's process would still act as a constraint. In the House of Representatives, the majority party controls both the process and substance of legislation, akin to the Canadian Parliament in majority government circumstances; however, the Senate procedure is governed by rules that have evolved to create a higher standard for advancing legislation, such that most bills require sixty votes, typically inclusive of votes from the minority. Republicans controlled fifty Senate seats, so conventional legislation would not pass any climate provisions. The only workaround is the process known as reconciliation, originally intended to advance budgetary resolutions with only fifty votes. In 2021, then, any federal climate legislation would have to be consistent with that budgetary standard. So instead of direct mandates, regulations, and prohibitions, fiscal incentives would be the form of climate legislation to conform to reconciliation.

The Democrats could have attempted to change the rules of the Senate, to dispense with the sixty-vote standard by removing the Senate practice of filibuster — the open-ended right of any senator to debate — but Joe Biden was not prepared to take that step; at least, not in 2021. Recall that he had spent thirty-six years in the Senate, beginning in 1972, when he was first elected from Delaware. Over that time he spent some years in the minority, and relied on Senate rules for obstruction via filibuster to create time for rational persuasion and to drive bipartisan consensus. Moreover, Biden had never been intensely ideological; rather, he was more known for reliably defending classic Democratic bedrock principles, from abortion to social security to union rights. That said, he had certainly seen the Senate evolve by 2008 to be more a partisan entity, more parliamentary and less collegial.

Biden distilled the GND's ambitions into a tangible legislative objective to decarbonize the American electric generation sector by 2035, which he deemed the most stringent proposition that could realistically get through the fifty-fifty Senate and the reconciliation process. In August 2021, the Biden administration achieved a bipartisan agreement with sixteen Republican senators on legislation that would improve physical infrastructure in the United States. The legislation would cost the country about $1.2 trillion; for context, the U.S. federal debt had reached almost $26 trillion, and its budgetary deficit for 2021 was expected to reach almost $3 trillion. A major component of the infrastructure bill would improve the U.S.

electric grid, making it more reliable and more compatible with renewables. However, it contained no other direct climate provisions.

In late August 2021 the House of Representatives disclosed a companion bill to the infrastructure plan. This bill dealt both with climate and improving the country's social safety net. Some had hoped it would contain a conventional clean energy standard akin to California's, mandating a decarbonized electric generation sector by 2035 — but it did not. To accommodate the constraints of reconciliation, the bill instead included provisions known as the Clean Electricity Payment Program (CEPP), a system for rewarding utilities that voluntarily met low-carbon requirements and penalizing those that did not. The carrot-and-stick approach to prioritizing wind and solar over coal and gas was meant to double the amount of wind and solar on the market, moving the nation's electricity sources toward 80 percent renewable by 2030, and even 100 percent by 2035. Other climate provisions dealt with methane emissions, tax credits for electric vehicles, building retrofits, and technology development, notably CCS. These climate provisions were broadly accepted as sufficiently credible to satisfy most of the practical environmental expectations. Furthermore, if passed, this legislation would tangibly demonstrate to the world that the United States was more seriously committed to decarbonization.

By October these provisions had been substantially altered, however. West Virginia senator Joe Manchin, who was elected for a second Senate term in 2018 by a slim margin of 49.57 to 46.26 percent, had refused to support the CEPP due to its impact on coal and natural gas; nominally a Democrat, Manchin represents a state that depends on hydrocarbon production, especially coal. Manchin is in the coal business, too; he founded a coal brokerage in 1988, although he has had no role in its operations since he committed to a political career full time in 2000. Based on his financial disclosure requirements for 2020, his continuing financial interest in that business represented only 31 percent of his net worth.

As a Democrat, he was committed to expanding the social safety net and was willing to add to the deficit, but his generally modest left-of-centre political position did not extend to provisions for explicit decarbonization. With fifty Republicans opposed to the $3.5 trillion reconciliation package that included these climate provisions, the bill could not move forward without Manchin's vote. Manchin, de facto, held a veto on how extreme any climate legislation would be.

Biden attended the Glasgow climate meeting of November 2021 with no federal legislation in hand, only his commitment to the aspirational goal of net-zero emissions by 2050 and an increased American emissions reduction target of 50 percent below 2005 by 2030 — despite that, as of 2020, U.S. emissions were only about 22 percent lower than 2005 levels.

At the end of 2021 the Biden administration was still working to find a formulation of effective climate legislation that Manchin could abide in lieu of CEPP, specifically trying to go forward with $500 billion in tax credits for low-carbon initiatives such as renewables, retrofits, electric cars, CCS, and the like. Of course, meeting Manchin's demands might not be acceptable to various climate advocates in the Democratic caucus in the House. Stalemate was as likely as even a modified breakthrough.

## The Three Energy Crises and Extreme Weather of 2021

Throughout 2021 the world prepared to descend on Glasgow to commit itself tangibly to net-zero emissions; meanwhile, a series of events in Texas, Western Europe, and China only reinforced how difficult any such energy transition would be. In February North Texas experienced its coldest temperatures in seventy-two years — -2°F (-19°C) at Dallas Fort Worth International Airport on February 16. The state's electric generation system and natural gas supply infrastructure had not been adequately winterized to deal with such conditions, and the resulting economic losses exceeded $130 billion. The system operator attempted to ensure that the system remained functional at the peak of imbalance between supply and demand so that power to essential services was not lost, and subsequent rationing of available supply made the ordeal more difficult for non-essential businesses and for households. Electric supply contributions from the state's renewable sources were lost as well, and disruptions of natural gas supply extended beyond electrical supply to basic space heating.

Recriminations abounded. Did Texas have inadequate regulatory oversight of its entire electrical generation system? Should regulators have demanded winterization sufficient to deal with an unlikely but still possible weather event? How much did renewables capacity exacerbate the situation? What was the right level of supply "insurance"? Should Texas have integrated better with other states' electric systems to reduce imbalances of supply and demand? Some posed the question of whether climate change was the root cause of the extreme cold snap itself. According to some climatologists, climate change could have made this extreme event more likely by displacing the polar vortex southward. In any case, the experts agreed that more must be spent, going forward, to diminish the impacts of extreme weather events like this one.

In 2020 wind power in Texas had ascended to almost 30 percent of installed capacity; and though its expected contribution for February 2021 was only 7 percent, due to diminished seasonal wind, many on the right unreasonably demonized renewables for causing the freeze's extreme impacts. In fact, the base load supply, still provided by fossil fuels, should have been winterized better to ensure continued operations during extreme weather conditions. Texas knew how to winterize, but in 2020 and 2021 wasn't willing to pay for it because most years that expense would have proved unnecessary. More renewables in the Texas electric generation capacity mix arguably would have made conditions more difficult, depending on how backup supply was installed and maintained from either natural gas or battery storage.

In fall 2021 Britain faced its own energy crisis. By late October European markets saw natural gas prices increase in spot trading by over 250 percent of traditional values, with prices approaching US$30/MMBtu (Metric Million British Thermal Unit). (For context, North American natural gas prices had remained, for most of the previous decade, below $3/MMBtu, even through extreme winter, thanks to abundant continental supply.) Electricity prices escalated comparably. North Sea drilling had slowed, and onshore fracking had been banned in Britain over the prior decade. Undue reliance on wind energy proved a mistake when Britain had its least windy summer in sixty years. As availability of North Sea wind energy diminished, production dropped to 7 percent of Britain's energy makeup, and international gas prices soared. Efforts to maintain price caps proved unsustainable. Meanwhile, in Asia, economic activity started up again post-Covid-19, and resulting increases in LNG demand diverted such supply away from Western Europe even more. Diminished exports of Russian gas into Western Europe, inadequate investment in storage, and lost coal generation capacity exacerbated conditions. And the full brunt of winter was around the corner. This kind of crisis is a predictable manifestation of energy transition from fossil fuels. As the winter of 2021–22 unfolds, Britain and Europe could only wonder how high the final bill for basic energy might be. This is the absurdity of trying to reduce dependence on natural gas with no viable substitutes.

In 2021 China decided to keep its domestic electricity and space heating prices affordable simply by abandoning any constraints on supplying or using internally sourced coal to meet those demands. Existing coal mines operated at full capacity again, and new mines were approved. Coal consumption could increase without limitation through 2025. This represented a massive retrenchment of Beijing's earlier rhetoric about tougher environmental targets and sharp cuts to coal-fired power generation. The country even countenanced increased coal imports from Australia, a country some of whose imports China has restricted for political and diplomatic reasons. Industrial output would not be sacrificed, despite the Glasgow conference just around the corner. Electricity supply from other domestic sources diminished, exacerbating the problem. Less rain in southwestern China meant hydroelectric dams generated less power, and calmer weather in northeastern China meant that its wind farms generated less, too. No starker example exists of the trade-off between energy transition and sustaining current economic output. China had hoped to achieve emissions decline by 2030 and carbon neutrality by 2060.

These "energy crises" came in the context of a 2021 with various extreme weather events, most notably:

- Exceptional heatwaves affecting western North America during June and July, with many places breaking station records by 4°C to 6°C and hundreds of heat-related deaths. Lytton, in south-central British Columbia, reached

49.6°C on June 29, breaking the previous Canadian national record by 4.6°C. The town was devastated by fire the next day.

- Numerous major wildfires broke out. The Dixie fire in northern California, which started on July 13, burned about 390,000 hectares by October 7, making it the largest single fire on record in California.
- Western Europe experienced some of its most severe flooding on record in mid-July. Parts of Germany and Belgium received 100 to 150 mm over a wide area on July 14–15 over already saturated ground, causing flooding and landslides and more than 200 deaths.
- Extreme rainfall hit Henan Province of China from July 17 to 21. On July 20, the city of Zhengzhou received 201.9 mm of rainfall in one hour (a Chinese national record) and 720 mm during whole the event, more than its annual average.
- Twenty weather/climate disaster events, with losses exceeding US$1 billion each, affected the United States, according to NOAA, the U.S. National Oceanic and Atmospheric Administration.

According to initial studies carried out by NOAA that focused on the causes of the heatwave in the Pacific Northwest, such events are "still rare or very rare in today's climate but would have been virtually impossible without climate change." For the Western European flooding, comparable studies opined that the heavy rainfall "had been made more likely by climate change."

## From the 2021 Canadian Federal Election to the Glasgow Climate Summit

In July 2021, in response to the coming Glasgow Climate Conference in November, Canada increased its emissions reduction target from 30 percent to 40 to 45 percent of 2005 levels, while the Liberal government remained aspirationally committed to net-zero emissions. Nothing, however, changed policy wise in terms of new specifics. In mid-August Trudeau called an election for September 20 — an opportunity for serious debate on the course of Canadian climate policy. The Conservatives, under the new leadership of Ontarian Erin O'Toole, modified their position materially, accepting of carbon pricing, applicable to all Canadian emissions via a carbon tax, as their pre-eminent policy instrument. This represented a legitimate change in the Conservative Party's historic resistance to carbon pricing, but they advocated a significantly lower level than that proposed by the Liberals — only $50/tonne by 2030 on non-industrial domestic emissions, and for industrial emissions conditionally up to $170/tonne, depending on competitiveness impacts. O'Toole cited the Paris target as his party's fundamental climate objective but did not explicitly commit to net-zero emissions by 2050. His position would not likely generate the required emissions reduction to achieve even the 30 percent national reduction target.

The election virtually replicated the results of the 2019 election: a Liberal minority with the NDP holding the balance of power. The Conservative popular vote was the highest of any single Canadian party's but comprised only 34 percent of the electorate. Including votes for the other right-wing party, only 40 percent of the Canadian electorate supported a position not aligned with net-zero emissions, at least as a national aspiration. That said, the fundamental merits of the respective policy position were hardly debated adequately over the course of the campaign. The election was generally interpreted to affirm both the Liberals' management of the pandemic and their intention to "build back better," based on Great Reset principles, with special emphasis on economic opportunities inherent in the energy transition to net-zero emissions. Although no doubt disappointed by their failure to win a majority government, the Liberals could take heart that Canada was still a centre-left country, and that the self-identified socialist alternative held the balance of power — a party less distinct from the Liberals than ever before, certainly, at least, in respect of climate aspiration.

The election results reflected an extreme polarization of Canadian politics from region to region and between urban and rural Canada, with Alberta remaining especially misaligned with the political agenda laid out by the Trudeau Liberals. This reached new extremes a few weeks later, when Trudeau announced the new federal Cabinet and appointed Steven Guilbeault, an environmental activist from Quebec. This key member of the Canadian environmental establishment, implacably opposed to hydrocarbon development in the country, was now environment minister. The prior environment minister, Jonathan Wilkinson, would take over the energy portfolio. The result was that the two portfolios could operate more seamlessly than ever, with no unwanted advocacy or empathy for hydrocarbon production. If Catherine McKenna had been plagued by the country's ambivalent commitment to climate, especially in terms of specific interventions to achieve expected emission reductions, Guilbeault's appointment would presumably diminish any such concerns.

Trudeau and Guilbeault's intentions could not have been clearer. Hours after being sworn in, Guilbeault eagerly announced, "We will ensure through legislation and regulation — something we will need to develop — that emissions from oil and gas in Canada are capped at current levels and diminish over time." No other sector of the economy was singled out, only Alberta's hydrocarbon production industry. It was expected to atrophy in the cause of Canada achieving net-zero emissions. Guilbeault laid out this vision with no conditions — not even whether the rest of the world fails to carry out the same level of mitigation ambition, or whether Canada can actually afford to lose its growth opportunities in hydrocarbon production in a current world, where demand for hydrocarbons is still growing, and many other countries will supply them regardless of Canada's sacrifice. Similarly, Guilbeault did not clarify whether any price was too high when it came to reducing carbon emissions.

His statement was an astonishing insult to Alberta. Still, Guilbeault's appointment did represent some long-awaited intellectual honesty, signalling a commitment to action, not just promise and rhetoric. A government planning to embrace net-zero emissions finally had an environment minister with no illusions or hesitations about the full implications of that position, which meant eliminating fossil-fuelled electricity, transportation, and heat, as well as completely transforming agricultural methods. Regardless of the cost to Canada and especially Alberta.

Trudeau's democratic mandate to achieve its 2030 emissions reduction goal by whatever means necessary, regardless of any other apparent consideration, had been validated by the nearly 60 percent of Canadians who voted for the parties aligned with him, and no Liberal broke ranks over the target, which would remain the same if not made more extreme. None of the Liberal ranks resisted the Guilbeault appointment, not even those elected from Alberta, or those from Alberta, such as Finance Minister Chrystia Freeland.

In early 2022 the federal Conservatives removed Erin O'Toole as leader by a vote of their parliamentary caucus. The reason was fundamental dissatisfaction with his authenticity as a conservative, coupled with electoral failure. Would his successor accept the necessity of carbon pricing as indispensable to any credible Canadian climate policy? A genuine "climate denier" could ascend to the leadership, given the conservative electorate's broad antipathy to the costs imposed on Canada from hydrocarbon production obstruction in the name of climate policy. It would be a challenge for any successor to retain that core electorate while maintaining credibility on climate policy. It can be done, though. The policy elements I lay out in chapter 9 provide the basis for accomplishing that.

## Glasgow COP26: "Keeping the Dream Alive" — Just Barely

COP26 stretched out beyond its original two-week schedule, which came as no surprise. Its tangible results had been debated long before they became official, and remained contentious, but it did manage to salvage an agreement amongst all the attending countries, however inadequate for some, thereby ensuring that the UN process would continue.

Some specifics: Canada raised its emissions reductions target to 40 percent reduction by 2030 from 2005 levels; the previous target was 30 percent. The United States committed to a 50 percent reduction by 2030 from 2004 levels. Glasgow also saw bilateral and multilateral agreements to reduce methane emissions, phase out coal, and financially abet realizing net-zero

GHG emissions. India pledged to achieve net-zero emissions by 2070, while China made no change in its position from Paris, which is to say it has no emissions reduction targets over the remainder of the decade.

The NGO Carbon Watch consolidated four major analyses of the pledges and promises made at Glasgow. Their findings:

- If the developed countries do not meet their Glasgow targets, global temperature is expected to increase 2.6 to 2.7°C, with an uncertainty band of 2 to 3.6°C by midcentury.
- If countries meet both their conditional and unconditional nationally determined contributions, the formal term for their near-term targets for 2030, projected warming is expected to fall by 2100 to 2.4°C, with a 1.8 to 3.3°C uncertainty range.
- Temperatures will likely peak in the middle of this century at 1.9°C if all countries meet their existing targets and honour their long-term net-zero promises.

The IEA offered a more optimistic outlook, taking solace in the pledges of seventy-four participating countries to reach net-zero emissions. The IEA's analysis found that these pledges, if met, would result in warming under 2°C by 2100. However, to limit warming to 1.5°C, global emissions would need to fall roughly in half by 2030. United Nations Energy Program analysis validates that revised emissions reduction targets would modestly reduce warming outcomes and projected 2030 emissions, but a massive gap still remains to keep 1.5 even as a possibility without resorting to removing part of the existing accumulation of GHGs in the atmosphere, an endeavour that poses formidable technological, governance, and sustainability risks.

Over the course of the summit, there was no serious discussion or analysis of the cost of "keeping 1.5°C alive," nor whether the developed countries could afford to make the required energy transition; and nor, for that matter, if they could afford the adaptation and damages funds promised to developed nations back at Paris and Copenhagen, on the order of at least

US$100 billion annually — let alone the more recent demands of trillions annually. No one emphasized that electorates of developed nations will be expected to fund these wealth transfers, even as their basic energy supply becomes more costly and insecure. Finally, of course, skeptics asked whether developed countries would actually act to meet the reduction targets they set in Glasgow. The gap between promise and delivery has been endemic to the UN climate process. They continue to ask, and fairly so: What government has the political will to impose the necessary costs on its citizens?

What Glasgow achieved was deficient, certainly by its own expectations. Its closing declaration included no reference to "phasing out" coal, let alone any other hydrocarbon. Nevertheless, countries were expected to meet again in 2022, at COP27, with even greater emissions reduction targets. Meanwhile, on December 19, 2021, Senator Manchin stated unequivocally that he would reject Biden's legislation containing the remaining elements of his administration's climate agenda — roughly $500 billion in fiscal incentives for various low-carbon technologies. The possibility remained, going into 2022, that some elements may be resurrected in new legislation, but Manchin, who chairs the Senate Committee on Energy and Natural Resources, stated,

> If enacted, the bill will also risk the reliability of our electric grid and increase our dependence on foreign supply chains. The energy transition my colleagues seek is already well under way in the United States of America.... In the last two years, as Chairman of the Senate Energy and Natural Resources Committee and with bipartisan support, we have invested billions of dollars into clean energy technologies so we can continue to lead the world in reducing emissions through innovation. But to do so at a rate that is faster than technology or the markets allow will have catastrophic consequences for the American people like we have seen in both Texas and California in the last two years.

Manchin made no mention of carbon pricing, carbon budgets, Glasgow, or net-zero emissions — and ended the year with a situation emblematic of the United States' evenly divided polity. Every Congressional Republican would likely reject the notion of a carbon tax out of hand, largely for reasons having nothing to do with climate change risk reduction and largely due to their fundamental antipathy to any incremental taxation, and perhaps even based on an unjustifiable attachment to climate change denialism. In any case, in the last hours of 2021 Manchin killed Build Back Better, and the Biden administration offered nothing in respect of carbon pricing, either.

## Energy Geopolitics Rediscovered: Merkel's Failure

Early in 2022, for the first time since the end of the Second World War, a major land war erupted in Europe. Russia, under Vladimir Putin, insisted on restoring Ukraine to Russian control, ending its seeming inexorable integration with the rest of Europe, both economically and in its defence arrangements. This was a sad consequence of mounting tensions over the dismal equivocations of the newly elected German coalition on the question of resisting Russian assaults on Ukrainian sovereignty. Germany's position was largely the result of its utter dependence on Russian natural gas imports to literally to keep its economy running. The nearly operational Nord Stream 2 pipeline would further cement this dependency, adding more overall import capacity into Germany from Russia, and also following a route across the Baltic Sea that bypasses existing land routes through Ukraine and Poland, increasing Russian control.

Germany's commitment to Nord Stream 2 stands in stark contrast to Angela Merkel's much proclaimed *Energiewende*, her grand plan to transform its energy supply to make it "low-carbon, environmentally sound, reliable and affordable." She planned to shift away from fossil fuels and nuclear generation in favour of a massive push toward renewable energy, primarily wind power, and she established bold emissions reduction targets, committing to a 40 percent reduction in greenhouse gas emissions compared to 1990 levels by 2020, 55 percent by 2030, and 95 percent by 2050, with renewable energy capacity to account for 80 percent of Germany's primary energy mix by the middle of the century.

Despite the billions that Berlin spent to effect this de facto decarbonization energy transition, Germany was on course to miss its first milestone emissions reduction target by some margin in 2020, with longer-term goals even more implausible. A major contributor to that outcome was the absurd decision to shut down German nuclear capacity after the Fukushima incident in 2012, roughly the same time period that the *Energiewende* was imposed. Perversely, removing low-carbon nuclear from the mix has effectively extended the operations of German coal-fired plants, many of which are fuelled by lignite, soft coal that is easier to mine but far more carbon-intensive to burn. And predictably, wind power has more been more problematic to install than originally anticipated: again, both cost and NIMBYism have played a role in limiting its adoption.

Increasing Russian gas imports became a last resort to deal with the failures of Germany's *Energiewende*, but these have come at the woeful price of greater dependency on Putin's regime. Maintaining nuclear capacity, developing longer term alternative LNG supply options from North America, and revisiting extreme emissions reduction targets were all eschewed. At the beginning of 2022 it was not incredible to think that Germany would substantially acquiesce to the Russian invasion of Ukraine.

On February 24 President Vladimir Putin of Russia decided to carry out a "special military operation" in Ukraine. In truth, Putin was launching a full-scale invasion, acting on his long-standing and visceral intention to restore Russian control over this sovereign nation — a clear violation of the presumed basic principle of international law, that international borders are not to be changed by force. The invasion was as unjustified as any since the Second World War, comparable to Hitler's invasion of Poland. Of course, the West, led by successive U.S. administrations going back to 2000, and coupled with indifference from major NATO members such as Germany, had let Russian forces build up, focusing on ineffectual diplomatic efforts, holding back on serious sanctions, and failing to send more military resources and expertise into Ukraine, air support in particular. In this failure Biden was abetted substantially by the irresolution of NATO members, most notably Germany.

As the invasion unfolded, world oil prices escalated well above US$100/bbl by early March 2022. Despite the West's imposition of severe economic sanctions on Russia shortly after the invasion began, Russian natural gas and crude oil exports into Central Europe were not disrupted, reflecting the region's short-term dependence on Russian fuel. Europe would have been less vulnerable and compromised if it had eschewed this trade with Russia earlier and turned instead to its NATO allies, notably the United States and Canada, for greater natural gas supply. Europe had relied on the idea that it could sustain closer economic integration with Russia without facing exploitation by Putin — a massive strategic blunder. Europe's excessive renewable capacity had only made it more reliant on Russian gas, and, worse, the trade helped the Putin regime fund its military resources and continued to do so as Russia invaded Ukraine.

Over time, how will Europe, but Germany in particular, respond to its dependence on Russian natural gas, especially if there is no fundamental regime change in Russia itself? Will Europe diversify its supply of LNG and replace the oil and gas that it imports from Russia with fuel from other sources, especially North America? Or will countries such as Germany rediscover the necessity of utilizing the non-emissive option of nuclear power as a fundamental component of base-load electric generation? The way forward will also require abandoning long-standing delusions that we can run modern economies on windmills and batteries, and this is true for North America as much as it is for Germany.

In fact, if North America produces more natural gas and backs Russian gas out of the global market, methane emissions will materially decrease, by as much as or more than what could be achieved by any further mitigation excess imposed on North American production by the Biden administration or the Trudeau government. With the strict emission regulations and industry operating standards in place for hydrocarbon production and transportation, fuel from North America is significantly cleaner than the hydrocarbons produced in Russia. So, increased American and Canadian LNG export to Central Europe would be the moral equivalent of the 1948 airlift that salvaged West Berlin from Russian control.

# PART TWO

# Reconsidering Climate Policy

# CHAPTER 7

# Covid-19 Sacrifice as Predicate for Decarbonization?

## Accountability for Covid Strategy and Tactics

"War is too important to be left to the generals." So declared George Clemenceau, prime minister of France during the First World War, which killed 1.4 million French citizens out of a population of 39 million. As of January 2022, the United States attributed 886,000 deaths to Covid-19, out of a population of roughly 330 million, while Canada attributed 33,870 deaths to the virus, out of a population of 38 million. Was Covid-19 important enough to be left to the epidemiologists? Is climate change risk important enough to be left to the climate scientists?

Let me emphasize at the outset: some future mutation of Covid-19 may evade the capacity of existing vaccines to reduce its impact materially. As yet, however, in 2022, that has not occurred; at least, the virus has not proven an existential risk to humanity. However, at the outset of the pandemic, in March 2020, the risk was so uncertain that governments implemented extraordinary measures, causing massive economic contraction and removing basic human rights related to assembly, worship, and dissent. The electorates of the world's democracies overwhelmingly accepted these measures as justified. I fear the same kind of rationalization and extreme measures that made sense in that context will be imposed in the context of climate change — a phenomenon that may appear to be of the same kind if we take the current rhetoric literally, but that is actually, in vital ways, a different type of problem, one that requires a different solution.

Covid-19 and climate change have some commonalities. In both cases, independent, unconstrained actions by individuals, businesses, or states can create risk and collateral damage to others — refusing vaccinations in the case of Covid-19; consuming increasing amounts of hydrocarbons in the case of climate change. Without government intervention, both risks are likely to grow; they need to be reduced and, ideally, contained. But governments should be obliged to ensure that the cost of mitigating any risk is not greater than the cost of the risk itself, especially when the elimination of hydrocarbon use lies in the balance.

Hydrocarbons — fossil fuels — have played and continue to play a fundamental role in the material progress of humanity by providing cheap, abundant, reliable, effective, and safe energy and materials via petrochemical processing. Moreover, no equivalently economic substitutes exist, apart from those that may appear in specifically fortuitous circumstances. And yet, the rhetoric of the various elements of the environmental movement demands adopting the extreme measure of decarbonization, regardless of whether such a policy objective can meet any fundamental test of net benefit.

Elected leaders must confer with credentialled experts to a specific degree to reach optimal policy; if political leaders should "follow the science," it does not follow that they should delegate all authority or accountability to those experts. Rather, political decisions ought to be consistent with limits of the scientists' consensus. As Clemenceau's famous dictum implies, only duly elected politicians should be the final arbiters of policy, which in the end is subject to democratic validation. Some, unsettlingly, seem to question the right to dissent. Too often, however, other "elites," especially in the media but also those in the corporate, academic, and environmentalist spheres, assert a disproportionate influence, endorsing tactics motivated to serve objectives of their own — not to deal with actual risk at hand by means of a dispassionate analysis of cost and benefit. Debate over what legitimately constitutes following the science, how to maximize human welfare, or how to value "equity" over "efficiency" should not end even with decisions informed by the edicts of "the credentialled." When we start condemning dissent as disinformation and stifling debate over what constitutes optimal policy, we have begun our descent down a slippery slope to frightening and unacceptable constraints on free speech.

Dissent around Covid-19, and how it should or should not be tolerated, has been particularly problematic, and I say this despite being an ardent advocate of vaccinating populations against the virus. No other alternative exists that can lessen the risk as materially, and the more people are vaccinated, the more risk is reduced. Indeed, I find it bizarre that anyone questions this proposition. Admittedly, the vaccines do not eliminate risk from Covid-19, even before the advent of the Omicron variant, but any intellectually honest cost/benefit analysis would validate vaccination as the rational personal decision and appropriate for governments to prioritize. Even this clarity, however, does not justify stifling public debate on related policy questions by demonizing other views or even questions as equivalent to disinformation. Here, I refer to policy questions around vaccine mandates, ostracizing the unvaccinated, rationing health-care capacity, continuing lockdowns, facilitating treatment experimentation, and so on.

The precedent struck by Covid-19 is thus unsettling and ominous, especially if it is applied to climate change. And I fully expect that to occur, especially within Canada, where the population has historically tolerated its government compromising individual rights much more so than its American counterpart. I will clarify this concern by laying out how both Canada and the United States dealt with Covid-19 over late 2020 and throughout 2021, even as the world embraced decarbonization to deal with another presumed existential risk.

## Finding Equilibrium with Covid-19, Mid-2020 to the End of 2021

By the end of March 2020, virtually the entire developed world was locked down, some countries and jurisdictions more severely and effectively than others. Billions of people voluntarily accepted restricted freedom of movement and access in order to minimize the spread of Covid-19, hopefully to buy time for public health officials, and most importantly to avoid overwhelming hospital capacity. Businesses not deemed essential attempted to continue operating online, while some outdoor economic activity in construction, agriculture, and food supply chains continued, implementing special conditions to deal with Covid-19 transmission. Schooling for children took place virtually in most jurisdictions, despite the collateral impacts on teaching efficacy and socialization. Personal outdoor activity was constrained to essential activities, primarily accessing food supplies and medical services.

A month into the pandemic, in April, the International Monetary Fund declared that world faced the worst economic downturn in global economies since the Great Depression, though this time it was one created entirely by governments' resort to lockdowns as their major tactic to contain the spread of Covid-19, an effort supplemented by travel restrictions, quarantine policy for those crossing international borders, social distancing, and masking. This was an economic contraction dictated by government and, even more to the point, by public health officials. The public overwhelmingly complied, at least initially. Developed countries did not know how long they could sustain their lockdowns economically, even as those mandates persisted in some form for most of 2020.

On May 15, 2020, the Trump administration announced Operation Warp Speed, a public-private partnership to create an effective vaccine by the end of 2020, using an infusion of nearly $10 billion of public funding. The initiative evoked much skepticism from Trump's usual political opponents and from the public health-care establishment.

Over the ensuing summer of 2020, cases fell and some North American jurisdictions cautiously reopened, many only to reimpose restrictions in the fall as new Covid-19 variants asserted themselves, causing cases to increase. Over the remaining months of 2020 promising vaccines were in their final clinical trials, and world financial markets approached pre-pandemic levels, despite having reached near panic levels back in late March 2020. Schools struggled to reopen in the fall, even as teacher unions resisted sending teachers back into the classroom. The future looked much brighter when, in November, Pfizer-BioNTech announced that its vaccine tested as 90 percent effective, and Moderna announced its comparably effective vaccine shortly afterward. In the middle of December both vaccines became available in the United States, by emergency authorization.

In 2020 the American economy shrank by 3.4 percent, the country suffering its biggest decline in GDP since 1946. The Canadian economy shrank by 5.3 percent, and Alberta's economy suffered more because of the impact on its resource sector, shrinking by 8 percent. However, in the second half of 2020 economic growth in the United States and Canada turned positive, thanks in part to massive stimulus spending by their respective federal governments, less severe lockdowns, and the prospect of effective vaccines in 2021. Such spending was essential to offset the impacts — primarily, lost employment — on those economic sectors that could not operate due to the various lockdowns implemented. The economic cost of Covid-19, including government-mandated economic contraction and debt financing of stimulus, was on the order of US$16 trillion, as estimated by former U.S. treasury secretary Lawrence Summers in late 2020, well before anyone knew that the pandemic would last into 2022. For context, extreme weather events in the United States in 2021 were valued at US$145 billion, according to the NOAA.

Governments established the basic pattern of dealing with Covid-19 in 2020. When rising case counts threatened to lead inevitably to higher hospitalization, ICU use, and ultimately deaths, governments imposed restrictions. This persisted into 2021 — jurisdictions let ICU capacity drive public health policy and the severity of restrictions, regardless of the aggregate level of vaccinations. ICU capacity, especially in publicly funded systems, were not designed to cope with pandemic conditions and still meet normal-course health demands. The expectation was that vaccinations would, over time, reduce cases and allow easing of restrictions, until a variant asserted itself to once again increase cases and threaten to overwhelm hospitals, particularly ICUs.

Over the winter of 2020 and 2021, initial variants of the original virus caused daily Covid-19 cases in the United States to exceed those from the first and second quarters of 2020. By the end of March 2021, vaccinations in the United States exceeded eighty-two million, with the most vulnerable vaccinated first. Initially, Canadian vaccination rates lagged behind the United States because of initial constraints on vaccine procurement. But the process moved on inexorably in both countries for the majority of their populations. Meanwhile, the United States had substantially opened up by the end of the first half of 2021, and the Biden presidency even mused about having vanquished Covid-19. Canada was more cautious, reflecting in part its lower vaccination rates over the first half of 2021; however, the advent of the Delta variant in North America, in early August, saw certain restrictions reimposed in various jurisdictions. Still, there was no full restoration of the kind of lockdowns that typified much of 2020.

By November 2021 vaccination rates in Canada and the United States approached 80 percent nationally, although parts of the United States had considerably lower rates, in some cases barely over 50 percent. In the latter half of 2021 new cases of Covid-19 were overwhelmingly noted in people who chose to go unvaccinated, except for certain "breakthrough" cases, in which previously vaccinated people tested positive. Typically, these patients had mild symptoms and substantially lower risk of death, on the order of twelve times less than those who were infected and not fully vaccinated. Nevertheless, by the fourth quarter of 2021 the United States still had the highest rate of death per capita in the world, at 2,278 per million; by year end, almost 800,000 Americans, or 0.25 percent of the population, had died of Covid-19, the majority of those during Biden's presidency.[7]

The second highest death rate occurred in Britain, at 2,107 per million, with Germany third, at 1,159. Canada's Covid-19 death rate was, at that time, 766 per million. In China, fewer than 10 people per million had died of Covid-19 — a testament to the country's extreme containment tactics.

During 2020 and well into 2021, savings rates grew significantly in both the United States and Canada, in part due to fewer outlets for spending, online shopping notwithstanding, and to people simply saving more prudently due to the uncertainty of the pandemic. At the same time, many households were financially distressed, struggling to pay for necessities such as rent and groceries — a cruel irony. One estimate suggests that Americans added nearly $4 trillion to their savings, but most of the gains went to the already wealthy. That same phenomenon occurred in Canada; in fact, Canada was estimated the greatest saver, in relative terms, of all G7 countries; but again, as in the United States, the highest income Canadians achieved the greatest savings. Rising stock markets over the course of the pandemic added to the net wealth of that same class in both countries as well, and wealth inequality actually grew over the course of the pandemic. Remote work obviously proved simpler and more advantageous for higher-income professionals than for those in service industries, which were drastically impaired.

Global economic growth for 2021 was projected on the order of 6 percent, in contrast to a 3.6 percent decline in 2020. This reflected, in part, how the world had managed to adapt to the reality of the virus. Obviously, the advent of vaccines was the overwhelming factor in this recovery and in restoring substantial normalcy for most vaccinated people. But approaching a 100 percent vaccination rate still proved elusive, as a residual element in virtually all developed economies rejected vaccination. Their reasons seemed inexplicable in the face of enormous empirical evidence for the vaccines' efficacy, but even when governments virtually ostracized the unvaccinated from many aspects of normal services, enough people resisted that Covid-19 cases swelled again in late 2021 from the Delta and Omicron variants. Various jurisdictions considered reapplying restrictions across the general population to avoiding breaching critical ICU capacity, the truly critically constraint in all jurisdictions.

Global GHG emissions as estimated by the IEA fell in 2020 by about 5.8 percent, or by almost two gigatonnes of $CO_2$, almost entirely attributable to Covid-19 impacts. However, 2021 saw a rebound of 5 percent in emissions, as demand for coal, oil, and gas were restored thanks to the global economic recovery. Emissions approached pre-pandemic levels, essentially thirty-three gigatonnes annually; for context, Canada's attributed emissions are the order of 0.7 to 0.8 gigatonnes annually. Crude oil prices in 2021 exceeded pre-pandemic levels, averaging close to $70/bbl. Natural gas pricing escalated comparably worldwide. This was in part due to restored demand, but also due to supply-side constraints, especially in North America, as discussed in part one.

At the end of 2021, few countries remained committed to eliminating Covid-19 from their populations entirely, though China still professed to be an exception. Even Prime Minister Jacinda Ardern of New Zealand, who had become an icon of complete Covid-19 containment, conceded that zero percent Covid-19 was no longer tenable. Her language was instructive. "With Delta," she said, "the return to zero is incredibly difficult, and our restrictions alone are not enough to achieve that quickly. In fact, for this outbreak, it's clear that long periods of restrictions have not got us to zero cases." Ardern was acknowledging that her government could not maintain isolation, quarantine, and lockdown while lagging behind the rest of the world on vaccination deployment. The country's former tactics were no longer politically, let alone economically, sustainable. The new norm across the developed world was to vaccinate as much as possible, and then live with some endemic Covid-19 risk, provided that hospital capacity is not unduly strained.

Prior to the advent of vaccinations, no jurisdiction had seriously contemplated, let alone tried, removing restrictions and letting the pandemic run its course, to the end of achieving herd immunity. In its most extreme formulation this approach would have entailed exposing the entire population to the virus, save for the most vulnerable. No government was prepared to accept a 1–2 percent death rate. Only because of the advent of vaccines — a genuine technology breakthrough — could many nations restore a large degree of normalcy and spare themselves the agony of trade-offs between endless lockdowns and higher overall death. Vaccination was virtually a silver bullet — full doses proved 90 percent effective at avoiding intervention, and even more effective at preventing death, without serious widespread side effects. Even in the face of the Delta variant that defined Covid-19 risk in the latter half of 2021, the vaccines performed. The cost of developing and distributing the vaccines had not even discernibly impacted the economies of the developed world. The advent of Omicron at the very end of 2021 added some doubt to the efficacy of current formulations and frequency of the doses, but initial assessments suggested that in a worst case, the vaccine could be adjusted. It became clear by March 2022 that existing vaccines, along with extensively applied third-dose boosters, held up against Omicron, at least in decreasing cases of severe illness. Much of North America and Europe began to end remaining restrictions and reopen.

○ ○ ○

The most important thing to be noted about the effects of the pandemic, for the context of this book, is the way in which the populations of developed countries behaved before and after the advent of the Covid-19. Pre-vaccine, populations overwhelmingly complied with their government's lockdowns and related mitigation tactics. There was some protest, but no extreme civil disobedience, and the police and military did not need to intervene, at least not in North America. That compliance was driven by a collective recognition that, given the uncertainties of this potentially lethal illness, minimizing contact was justified. Populations felt the urgency even more in jurisdictions that came close to breaching hospital capacity. Weighing risk against reward compelled people to willingly give up certain basic rights of assembly and choice, without knowing when or how the situation would resolve, and we deferred to the best judgments of certain experts in public health.

To achieve acceptable containment of the virus within a discrete jurisdiction, let alone nationally or globally, vaccination rates would need to approach, if not exceed, 90 percent — likely an impossible feat without governments ruthlessly applying mandates for total vaccinations, with virtually no exceptions. As of the end of 2021, no leaders had resorted to such mandates. The Biden administration tried to impose vaccination requirements on employers, but legal resistance was largely validated by the courts. In place of forced vaccination, jurisdictions including Canada and the more liberal U.S. states de facto ostracized the unvaccinated, barring them from a broad array of non-essential services. Jurisdictions agonized over imposing various restrictions in the face of rising case counts — restrictions that would primarily reduce risks to the unvaccinated minority. But as of the end of 2021, no jurisdiction was prepared to deny access of the unvaccinated to essential services, most importantly hospitals. No jurisdiction was willing to increase the cost of being unvaccinated, despite the cost imposed on the vaccinated majority. No developed country even mused on such actions. Admittedly, Quebec's Legault regime mused about increasing personal taxation on the unvaccinated, but that idea was soon dismissed.

In a liberal democracy, we have the right to choose the level of risk we expose ourselves to, provided that we impose no collateral risks on others. In the case of Covid-19 vaccination, going unvaccinated does impose such risks, not only because it risks exposing others to the virus, but because it crowds out access to hospital capacity. This latter point is especially problematic in a publicly funded system such as Canada's, since it already has to make difficult allocative decisions about access to medical services. But the unvaccinated were not denied access, nor were they required even to pay direct additional costs for that access. But governments have been prepared to impose costs on the unvaccinated by denying access to non-essential services, constraining personal rights of assembly, impacting their economic livelihoods in some cases, and so on.

Any conservative government faced a genuine dilemma between its ideological commitment to personal freedom and the inescapable fact that dealing with the Covid-19 pandemic, pre-vaccine, required collective action. In the post-vaccine era, overall risk is reduced ever more as the vaccination rate increases. One might expect those on the right to consider sustained economic recovery another compelling reason to vaccinate the entire population, a course that promises an inherent and overwhelmingly positive net benefit. But imposing vaccination remains too intrusive to contemplate seriously. The Kenney government in Alberta was a classic example. Eager to fully reopen in the summer of 2021 and motivated as much by legitimate economic concerns as civil rights, Kenney lifted most constraints and mandates in early July, allowing large gatherings with no masks or social distancing. Notably, and symbolically, the Calgary Stampede at the beginning of that month was business as usual. As the summer continued the Delta variant spread, and Alberta's vaccination rates had not yet reached 70 percent for one dose. By fall, Covid-19 levels had, of course, skyrocketed, and patients nearly overwhelmed the province's hospitals and ICUs. Grudgingly in September, Kenney restored not only masking and social distancing requirements, but imposed vaccination requirements for access to non-essential services in Alberta. Ideological concerns for individual rights were subordinated to maintaining hospital capacity. Fortunately for Alberta, rising vaccination rates and the restored restrictions kept the hospital system from being overwhelmed. ICU capacity, ultimately, becomes the benchmark against which virtually all jurisdictions measure their Covid-19 regulations — whatever is required to avoid breaching that capacity. Germany's and Britain's actions in late 2021 showed that this is true as much as Alberta's.

○ ○ ○

So, what does all this Covid-19 experience tell us about the path forward on climate policy and about the extremity of decarbonization? The governments of most developed countries in the world had accepted de facto, by the end of 2021, that zero Covid-19 was not a realistic or even responsible goal, in part because the economic and political costs would be too great to sustain. However, vaccinating virtually 100 percent of their eligible populations was achievable at relatively low incremental cost and would maximize potential resistance to the virus. Admittedly, 100 percent vaccination does not equate to zero Covid-19, but no better option existed for reducing the Covid-19 risk (notwithstanding China's continued application of methods more intrusive and ruthless than what any liberal democracy would, to date, countenance). By spring 2022 the application of vaccines had evolved to the point that Covid-19 became an acceptable risk, even in the context of the Omicron variant. Residual risk would still persist of the virus mutating to a form more lethal, more transmissible, and more resistant to existing vaccines. Again, vaccine technology can be expected to adapt in response, but the world also accepts that zero Covid-19 is not an objective humanity can reasonably pursue. We accept the residual risk. Can this reality inform living with a 3°C global temperature increase?

Was Covid-19 ever an existential risk to humanity? When the Delta variant became the dominant strain in America, unvaccinated Americans were eleven times more likely to die than those fully vaccinated, according to the U.S. CDC. We can infer from analysis that if the entire U.S. population had faced Delta without vaccines, the death rate from Covid-19 would likely have risen to eleven times the current nearly 800,000 deaths — that's just under 10 million people, in a country of 338 million. A horrific picture, to be sure, but one in which the vast majority survived, and one in line with original estimates of mortality rates on the order of 2 percent. To avoid that, would the United States have locked down indefinitely, regardless of the economic cost, to minimize overall deaths? Vaccines spared all developed countries coming to terms with that "Strangelove" trade-off.

The IPCC has never explicitly stated that reaching 3°C would mean human extinction, yet alarmism animates the UN process. Decarbonization advocates seemingly hold that no cost and no sacrifice is too great to contain 1.5°C — whether contraction of economic output or a return to pre-industrial living standards. Some still contend that we need not regress economically to achieve net-zero emissions, that transforming the world's energy systems to be non-emissive is technically feasible and economically tolerable, that the developed world, at least, can absorb the incremental costs of the transition. That is highly debatable in the context of existing technology to replace hydrocarbons, but the UN process has never spelled out the net cost of decarbonization and then compared that cost to the social cost of carbon, let alone advocated as a reasonable policy alternative uniform, transparent global pricing ascending over time to the level of that cost. This — namely, finding a more economically optimal global climate policy — should be the whole point of the UN process, but sadly is not.

In the context of Covid-19, the positive net benefit of vaccines relative to their cost is overwhelming and obvious. No serious person should dispute this after the empirical evidence provided during 2021, when the mass application of vaccines provided protection, if not immunity, from Covid-19 with demonstrable efficacy and efficiency, with no significant side effects. The risk of Covid-19 can be substantially reduced by an available, affordable, and material technological fix (that is, vaccines), but this is not yet the case for climate change.

o o o

Covid-19 poses an immediate risk to virtually all of humanity, while the negative effects of climate change will substantially be suffered by future generations, notwithstanding current extreme weather events. So, the benefits of our Covid-related sacrifices accrue to the same people who are making those sacrifices, while, in the case of climate change, those who impose decarbonization on themselves will incur massive costs in the short and medium term, for uncertain long-term benefits accruing in the future. Mobilizing existing populations to deal with Covid-19 is obviously simpler than imposing near-term costs on them for climate mitigation, especially when some of those costs are passed on to future generations in the form of increased debt, taken on to pay for lockdowns and the economic destruction that was an unavoidable result.

Intergenerational effects apply both to Covid-19 and climate. With Covid-19, debt financings to fund stimulus spending will be passed on over time. As for decarbonization, given the realities of existing technologies, current generations will bear most of the cost of energy transition, especially in any present value sense. But those fundamental costs to effect decarbonization will endure over time if not redressed by markets and democracies.

Complicating matters further is the fact that, as noted earlier, an asymmetry exists in respect of those countries that had virtually nothing to do with the accumulation of atmospheric GHGs but will still face some of the worst effects of climate change and with fewer resources to adapt. Dealing with that has bedevilled the UN climate process since its inception. Some transfer for adaptation must be a genuine component of any global climate policy, regardless of whether it is based on decarbonization or a 3°C equilibrium.

Both managing Covid-19 and dealing with the climate change require government intervention. In the case of climate change, the question of decarbonization should be very much an open question, but some intervention must occur, even if only to establish an enforceable and rational global carbon pricing system. For Covid-19, Western democracies have not directly forced vaccinations on unwilling citizens; instead, they have mostly taken indirect measures to increase the cost of being unvaccinated, essentially preventing people from accessing certain services, employment opportunities, and public events.

Throughout the course of the pandemic, courts, especially in Canada, have sanctioned virtually all actions taken by government to deal with the Covid-19 risk when those actions were contested in court. Minimizing overall death rates validated government interventions. And no courts insisted that governments table cost/benefit justification of their tactics on Covid-19 as much as they probably should have, if for only transparency. But government interventions in markets and compromising basic human rights that typify functioning democracies should always be of concern. I say that as a basic axiom of political reality. The "precautionary principle" can run amok. Invoking the rationalization that, but for our intervention, costs and deaths would have been worse than those incurred is always the recourse of the "credentialled."

The net cost benefits of Covid-19 vaccinations should be obvious to even the most ardent libertarians. Compromising personal freedom can be rationalized, and indeed occurs in all civilized societies, but an informed electorate still has the right, through the democratic process, to define how much such rights can be compromised and for what net benefit. For dealing with the climate risk, however, I have great unease that the extreme intervention of trying to achieve decarbonization will do more harm than good. Covid-19 is not a precedent that justifies decarbonization.

## Limits of Intervention: Covid-19, Early 2022

As 2021 closed, the advent of Omicron created as much anxiety as the world had experienced since March 2020. Full vaccination would still help to prevent infection or serious illness, and the basic mutation proved less lethal, certainly relative to Delta, but Omicron's greater transmissibility translated to more infections of those fully or partially vaccinated, despite likely lower severity. Breaching health-care capacity remained the determining factor for short-run tactics. Simply put, more cases could still potentially breach existing hospital capacity, and especially ICU capacity.

So, for the third or fourth time in virtually every Canadian province, some version of lockdown was restored, typically restraining access to non-essential services and assembly. Most of the United States did not reimpose actual lockdowns, even the most liberal states. Both countries were able, at least by January 2022, to restore in-person teaching for pre-kindergarten to grade 12, despite the expected resistance of teacher unions. Incredibly, most universities still insisted on virtual classes, at least over the first quarter of 2022. The Biden administration continued to call for higher levels of vaccination. The effort to require U.S. businesses of a certain scale to impose vaccination on their employees was struck down by the U.S. Supreme Court as beyond the presidential authority.

As of February 1, 2022, 63 percent of eligible Americans were vaccinated with at least two doses. In Canada, that metric was 82 percent. As of mid-January, in Canada, about 83 percent of all attributed Covid-19 deaths occurred amongst the unvaccinated or only partially vaccinated, while hospitalizations attributable to Omicron were at an all-time high. ICU utilization was at levels consistent with past peaks, still near 100 percent even as of the end of January. Cases attributable to Omicron are considered under-reported, but the trend in reported cases in Canada nevertheless trended downward from a mid-January peak of sixty thousand new cases per day down to about eleven thousand. Death rates remained on the order of 160 per day over the latter half of January.

Especially within Canada, the advent of Omicron exacerbated tensions over the tactics deployed to deal with it. Canadians noted that the United States, in contrast to Canadian provinces, especially Ontario and Quebec, did not resort to the lockdown tactics of early 2020, despite similar infection rates. Admittedly, ICU utilizations in the United States generally never approached the levels endured in Canada over the month of January — roughly 40 versus over 90 percent. This was perhaps as much an indictment of Canada's socialized medicine model in terms of available ICU capacity and other inherent rigidities that prevent increasing the supply of physical and human resources, Covid-19 notwithstanding.

More profoundly, over this first quarter of 2022, an unexpected development in Canada became emblematic of my great dread about what may yet come to pass if governments ruthlessly implement decarbonization. In late January, just as reported Omicron cases across the country had peaked and were beginning to decline, the Trudeau government declared that unvaccinated truck drivers crossing the U.S.-Canada border would no longer be exempt from Covid-19 vaccination requirements, an exemption they had held since the beginning of pandemic. Specifically, unvaccinated Canadian truckers would now be forced to quarantine for fourteen days upon re-entering Canada, while unvaccinated foreign truckers would not be allowed into Canada at all. The new mandate was in part a response to a reciprocal requirement that United States intended to impose by February.

Omicron was the only variant spreading in Canada at the beginning of 2022, transmissible to and by the vaccinated as well as the unvaccinated, though vaccination provided better resistance to serious impacts. The vaccination rate among those eligible was 90 percent, and that same rate held for truckers. Trudeau's government offered no detail on the marginal utility of imposing this mandate on the 10 percent of truckers who were not vaccinated, an estimated sixteen thousand individuals. Trudeau and his ministers offered no explanation for the new mandate, other than simply asserting that vaccinations were the best response available to lessen the impacts of Omicron, and that the federal government would therefore impose vaccination wherever it had jurisdiction to do so. They did not specify by how much Omicron impacts would be reduced by imposing a vaccination mandate on sixteen thousand truckers, relative to the supply chain disruptions caused by taking those workers out of service.

The visceral reaction to this federal mandate was unexpected and unsettling, not only for the government but also for those Canadians who had aligned with Trudeau politically, particularly in terms of managing the pandemic — his policy of minimizing attributed Covid-19 death rates regardless of the cost. Trucker protests erupted almost immediately. Thousands of drivers formed convoys and headed to Ottawa to demand that the government reverse the new mandate. The movement quickly received substantial funding from supporters in Canada and the United States. Truckers also blockaded border crossings in Alberta and Ontario, in solidarity. These protests attracted global media attention and became the *cause célèbre* of right-wing entities resisting government Covid-19 tactics across the developed world.

Predictably, Trudeau and the centre-left in general vilified the protesters as seditious extremists, unjustified in their demands and a threat to Canadian democracy. Mark Carney and Andrew Coyne both published opinion pieces in the *Globe and Mail*, articulating the Canadian Liberal position that the government must not engage with the protestors on the merits of the mandates, and certainly must not negotiate a settlement, but should use whatever force necessary to disband the protest, regardless of whether anyone was breaking the law. Fundamentally, they scorned the protesters' audacity to demand change from the Trudeau government. Carney wrote:

> The goals of the leadership of the so-called freedom convoy were clear from the start: to remove from power the government that Canadians elected less than six months ago. Their blatant treachery was dismissed as comic, which meant many didn't take them as seriously as they should have. Certainly not our public safety authorities, whose negotiations facilitated the convoy's entry into the heart of our capital and have watched as its dangerous infrastructure has been steadily reinforced — a policy of engagement that has amounted to a reality of appeasement.

In reality, far from engaging with them, Trudeau adamantly refused to meet with the protesters to seek accommodation, calling the conversation a "non-starter."

For the Canadian liberal media, the protesters were Canada's equivalent of Trump's base south of the border, assaulting not only specific Covid-19 mandates but the entire left-of-centre hegemony that has most consistently governed Canada — a hegemony validated in the 2021 election. The left-leading media framed the protesters' unwillingness to acquiesce as an "existential threat" to Canadian democracy. Canadian conservatives, on the other hand, held to a cautious line, decrying illegal acts among the protesters while at the same time criticizing Trudeau's failure to elucidate the actual merits of the vaccine mandate on sixteen thousand Canadian truckers. By the first week of February the police had made only minor arrests, and parked trucks occupied much of downtown Ottawa as well as certain border crossings. The stand-off continued.

The protests, at their core, were about vaccine mandates on unvaccinated truckers, and the situation was, as such, truly a collision of two stupidities. On the one hand, it was manifestly stupid, or at least ill-advised, to resist vaccination against Covid-19, especially when the unvaccinated were principally responsible for the strain on the hospitals and ICUs from Covid-19 patients — real third-party impacts. These individuals refused vaccination in a country with socialized medicine, where it was difficult to increase hospital capacity in the short run, and they were subject to no incremental costs if hospitalized with the virus.

On the other hand, however, it was misguided for Trudeau to ignore the actual costs and benefits of mandating vaccinations for the 10 percent of truckers who had been crossing the U.S. border without them, providing a genuinely essential service throughout the pandemic going back to March 2020. Trudeau also appeared oblivious to Omicron's differences from the other Covid-19 variants, and to the direct costs of trying to end the protests without genuinely engaging or negotiating. It was manifestly stupid not to engage on these matters, let alone offer a transparent account of the specific mandate's utility.

Sadly, the protests morphed beyond the specific merits of that one mandate, to focus more broadly on the government's licence to constrain personal action and choice in the name of a policy objective, even one as seemingly as laudable as containing the spread of Covid-19. That tension and conflict had been in play since the onset of the pandemic, and consistently proved a difficult balance. Compromising choice meant destroying economic activity, mostly unevenly and unfairly.

My grave concern, which I hope all serious and fair-minded people can admit, concerns the precedent Covid-19 mandates have set, and what mandates the government will apply in the future. The most obvious case, of course, relates to climate change. It should now be clear that the government with the same mentality on display during the pandemic could apply the same methodology to the end of decarbonizing.

Most leftist politicians believe a "climate emergency" is under way, and they argue that emergencies call for strict — sometimes ruthless — mandates. If the world is to be saved, sacrifices must be made. In fact, climate policy, if defined by the extremism of decarbonization, requires both massive and immediate intervention. Once any and all incremental carbon emissions are officially unacceptable — the inescapable logic of decarbonization, leaving aside the sophistry of net-zero — then governments can contend that their democratic mandates provide justification for almost any form of intervention, surveillance, and moral suasion. It is no coincidence that those who reacted most virulently against the trucker protests are those same Canadians so earnest in their advocacy of decarbonization. For climate zealots, the government's response to the trucker protests might be deemed to provide an instructive lesson for how to deal with other emergencies.

Predictably, as the trucker protests continued into mid-February, more extreme actions were implemented by various governments. The Ontario provincial government declared a state of emergency, creating more resources for policing of protests and penalties to be imposed on protestors, and a court injunction was concurrently obtained to legitimize police action at the Windsor-Detroit border crossing. But the protestors themselves largely stood down and withdrew from the border crossings in Windsor and in Coutts, Alberta, by February 15. However, the residual resistance remained in Ottawa.

On February 14, the Trudeau government announced in a press conference, not in the House of Commons, its intention to invoke, for the first time in Canadian history, the Emergencies Act. The rationalization? Trudeau stated, "It is now clear that there are serious challenges to law enforcement's ability to effectively enforce the law.... It is no longer a lawful protest or a disagreement over government policy. It is now an illegal occupation. It's time for people to go home." It is notable that Trudeau had rarely shown such moral clarity regarding the various unlawful blockades against pipeline construction.

Federal police forces could now intervene to enforce municipal bylaws and provincial offences where required, in lieu, presumably, of inadequate performance of existing policing authorities. But the real gambit was financial. It was left to the author and finance minister Chrystia Freeland to announce that Canadian financial institutions must temporarily cease providing financial services if the institution, or the Trudeau government, suspected an account was being used to further the illegal blockades and occupations. This order covered both personal and corporate accounts. As well, funding entities and platforms that supported the protests would be deemed subject to the provisions of the Proceeds of Crime and Terrorist Financing Act. What police could not accomplish on the ground would be accomplished by starving the protesters not just of funding but of access to their property, most notably their own bank accounts.

This application of the Emergencies Act went into effect immediately, but a vote in Parliament was required to sanction it. Not surprisingly, the NDP supported the Trudeau government, ensuring the measures came into effect, despite the opposition of the federal Conservatives and the Bloc Québécois. No real violence occurred throughout February, but no attempts at mass arrests in Ottawa had yet been made. Meanwhile, global attention on Canada intensified.

o o o

At the beginning of February 2022, Johns Hopkins University released a meta-analysis of the effects of Covid-19 lockdowns.[8] The findings were not salutary: during the first wave of the pandemics, lockdowns, the authors concluded, prevented only 0.2 percent of Covid-19 deaths, while reducing economic activity, raising unemployment, reducing schooling, causing political unrest, contributing to domestic violence, and undermining liberal democracy. Admittedly, this analysis was prepared by economists, not epidemiologists, but it showed that standard cost/benefit analysis leads to a strong conclusion: lockdowns should be rejected out of hand as a pandemic policy instrument. This report was a harbinger of intense debate over the efficacy of the tactics deployed to deal with the pandemic, especially in terms of their cost and benefit, and whether that consideration was ever part of the decision-making process at all.

The lesson here should be obvious. Before we contemplate, let alone implement "climate lockdowns" as a logical complement to the primary objective of decarbonization, we must analyze the costs and benefits and take seriously the measured words of Roger Pielke of the University of Colorado. About climate change, he states, "Apocalypse isn't on the table, and hoax isn't on the table." But climate lockdowns will have their advocates, especially among the likes of Mark Carney and Bill Gates. Imagine a world with limits on the distance that can be driven or flown, regardless of any carbon tax already in place; permanent virtual learning; a ban on gasoline and diesel vehicles, and on any device fuelled by hydrocarbons; surveillance to enforce carbon-footprint limitations; mandates on what foods are permissible; and mandated densification of living space. Orwellian enough? Truly, our personal freedom may soon be subordinated in this way, in the face of a presumed existential crisis. At the beginning of 2022 we saw the Canadian government invoke the Emergencies Act rather than reconsider a vaccination mandate of virtually no marginal utility. How, then, might it react to "code red for humanity"?

# CHAPTER 8

# Reconsidering Decarbonization: Not a Hoax . . . but Not an Apocalypse

Again, climate change is a serious global risk, but the optimal policies to deal with that risk should remain subject to debate and reconsideration. This proposition absolutely runs counter to the accumulated momentum of the UN climate process, and to the commitments that process has secured from the governments of the developed world. This is where things stood at the end of 2021, after the Glasgow Climate Conference; once again, the conference parties had further entrenched the goal of decarbonization, without genuinely confronting the net costs and benefits of that objective relative to other possibilities, such as adapting to 3°C. Again, in the context of climate change, adaptation involves making investments and taking actions that reduce the cost of the risk, but do not try to eliminate that risk. In the context of climate, mitigation involves taking actions to try to eliminate the risk, rather than merely trying to reduce the impact of climate change. For example, buying an air conditioner may help a person to adapt to higher temperatures, but the electricity that runs it must be from a non-emissive source for it to eliminate any associated GHG emissions. Of course, the latter costs more to accomplish.

Glasgow focused on targets and pledges, not enforcing their execution. Whether the politics and governmental structures of the developed world can actually deliver on their targets and pledges remained very much an open question. Will the costs of trying to deliver decarbonization prove affordable in the long term, economically and politically, in the developed world's functional democracies? Of course, China is no democracy, and if it does not align itself with global climate policy, then every other country's decarbonization efforts will be rendered futile. In China, the same question applies as everywhere else: Is the cost required to decarbonize ultimately acceptable? This skepticism about execution lies at the heart of most criticism of the achievements in Glasgow.

Make no mistake: dealing with climate change risk represents a major cost to the world, whether we decarbonize or adapt to a global temperature rise of 3°C above pre-industrial levels. The key questions should be how much net benefit can we achieve, if any, from any investment made or action taken, and who will pay for those? Global wealth distribution has been fundamental to the UN climate process from its inception, with North to South wealth transfers as much an objective as climate change mitigation. The key elements of Article 6 of the Paris Accord permit a country that has exceeded its Paris climate pledge to sell any overachievement to a nation that has fallen short of meeting its own emissions reduction obligations. This is relatively simple if the reduction obligation assigned to developing countries, or countries deemed to be developing, is minimal or even zero.

It is inescapably true that North America and Europe have emitted an overwhelming majority of the current of accumulations of GHGs in the atmosphere, notwithstanding the current high emission levels in China and India. It is also true that if GHGs continue to accumulate, and as certain climate impacts increase in frequency and severity, poorer countries will be impacted disproportionately and will have fewer, if any, resources for adapting. Poorer people everywhere, including those in Canada and United States, will suffer more from climate change effects than their richer compatriots. To state the obvious, those with wealth have more resources to deal with all risks, including those arising from climate change. However, pursuing decarbonization will erode living standards in the very economies expected to fund adaptation and mitigation for those vulnerable, both within their own borders and beyond.

The pre-eminent issue should be whether decarbonization is anywhere close to being the optimal policy for dealing with climate change risk, despite the IPCC's astonishingly hyperbolic contention, supposedly based on the four thousand pages of scientific reports from Working Group 1 of the Sixth Assessment Report, that climate change risk means "code red" for humanity. Can the reconsideration I describe come about, or is the momentum for decarbonization now irreversible?

## Polarization

People and groups on the left dismiss resistance to, or even skepticism about, decarbonization, owing as much to their fundamental political orientation as to any rational analysis of proposed climate policy in the context of reasonably projected low-probability, high-consequence events (tail risk) arising from climate change. Decarbonization requires not only massively changing fundamental energy systems, but massive and coordinated government intervention, whereby mandates supplant markets and "green becomes the new red."

Energy transition in response to climate risk, especially the extreme transition of decarbonization, can occur only by means of massive government intervention. The government must dictate what energy forms are exploited, and not only that, but how the energy produced can be consumed. This is the inevitable dead end of decarbonization: individuals will no longer enjoy choice in how and what they consume, since excess consumption would strain progress toward the goal. In a rapidly decarbonizing world, energy consumption must be allocated on the basis of criteria set by government, which almost inevitably will not countenance any policy revision in this area. If private property was anathema to classical Marxism, then the right to consume, and especially to consume hydrocarbons, is the private property of today's environmental movement. Prohibition of natural gas appliances or internal combustion engine vehicles will be the early manifestations of this inevitable descent into a green totalitarianism. Of course, transforming our energy systems provides a perfect synergistic moment for the fundamentally leftist agenda to deconstruct capitalist economies. And the great ultimate test, as it always is in any essentially totalitarian state, is whether political opposition will be allowed to persist, in this case on the merits of decarbonization. This is not hyperbole, but a legitimate scenario of the future, merely following the logic once we accept "code red for humanity."

One might reasonably question whether the likes of Al Gore, John Kerry, Bill Gates, Mark Carney, and the gatekeepers of McKinsey and their ilk are in fact ardent climate socialists rather than green capitalists who genuinely believe that technological breakthrough will soon bring an affordable, reliable, and green transition. No doubt most identify as the latter; however, I suspect they are deluded. Once decarbonization is actually embraced, no intervention is too much, and no constraint counterproductive. What is the practical difference between one constraint and another, once decarbonization ceases to be subject to any test of net cost and benefit?

Columbia University philosopher Adam Tooze, a respected left-leaning academic, is more honest about climate and capitalism when he writes that "any serious attempt at energy transition will involve, along with pricing and negotiation, a combination of nationalization, regulation, and prohibition, enforced not just according to the letter of the law, but with militant energy." Even more tactical is historian and climate activist Andreas Malm, author of the polemic *White Skin, Black Fuel* and most famous for his climate resistance manual, *How to Blow Up a Pipeline*, both published in 2021. In the latter, he writes:

> Here is what this movement of millions should do, for a start: announce and enforce the prohibition. Damage and destroy new $CO_2$-emitting devices. Put them out of commission, pick them apart, demolish them, burn them, blow them up. Let the capitalists who keep on investing in the fire know that their properties will be trashed. "We are the investment risk," runs a slogan from Ende Gelände, but the risk clearly needs to be higher than one or two days of

> interrupted production per year. "If we can't get a serious carbon tax from a corrupted Congress, we can impose a de facto one with our bodies," Bill McKibben has argued, but a carbon tax is so 2004. If we can't get a prohibition, we can impose a de facto one with our bodies and any other means necessary.

Later, Malm writes, "If nothing else, the anti-climate politics of the far right should shatter any remaining illusion that fossil fuels can be relinquished through some kind of smooth, reasoned transition.... A transition will happen through intense polarization and confrontation, or it will not happen at all."

Climate change has been a godsend for the left, because it offers a new rationalization for capitalist deconstruction. The dynamic is more Dostoyevsky's *The Devils* than even Pasternak's *Doctor Zhivago*. The thought leaders of the movement are distinct from Larry Fink and Mark Carney only in their intellectual honesty.

Ironically, there exists a link between the climate change goals of this group and those of their ideological opponents: those who would use decarbonization to advance their own self-interest, advocates who will prosper from mandated energy transition, including those who would finance it, those who would provide consulting services advising how to effect it, and technological infrastructure to enable the transition. Others, the ones who would police it, directly or indirectly, would make good from it, too. All of them become vested in the process, regardless of whether the extremism of decarbonization is justified by society at large. There are fortunes to be made from the trillions in new investments required to effect decarbonization, regardless of its net cost and benefit.

Left behind are the diminishing number of people — too many still, it must be acknowledged — who refuse to acknowledge that climate change is a real risk and must be dealt with. Theirs is a fundamentally untenable position.

In these times, polarization is not exclusive to climate change; indeed, it has never manifested itself in so significant, visceral, or consequential a manner as today, and that is true in virtually all public policy issues. In a world more transparent than ever, where information is freely available, and transmitting one's own opinion is simple and inexpensive, it is perhaps inevitable that populations would grow ever more polarized. The issue impacts all media, when journalists must consider how reporting a particular fact will affect political narratives. This is true of left- and right-leaning media, and the sad reality is that much of the media has transformed from reporting events to advocating outcomes and ideologies. Recent classic examples include binaries of riot vs. peaceful protest, self-defence vs. vigilantism, and meritocracy vs. white privilege. For a tangible example of the extremity of the political chasm, consider the divide between reporting and editorial positions in entities such as the *Wall Street Journal* and the *New York Times*, especially in respect of climate.

In my view, this polarization has at its root a fundamental disagreement on how to deal with socioeconomic inequality. To what extent do the existing economic and political systems in developed countries fairly deliver commensurate economic and social outcomes, individually, tribally, and nationally, and, if they don't, can they ever do so without radical fundamental change? The left has come to demand, if not to expect, not merely equal opportunities, but equal outcomes, regardless of effort or talent. It also demands that governments redress historical obstacles and injustices, regardless of whether any current taxpayers or wealth holders are responsible for them. All this is very much at odds with the concepts of individual liberty and personal accountability, and constitutes an assault on the long-accepted principle that holders of private property, whether real estate, cash, or other financial assets, may deploy those resources substantially as they choose. Of course, over the past hundred years, progressive taxation and regulation have increasingly fettered the rights of private capital — an uneasy balance resolved by the political process within long-standing democracies. But consensus on where that balance should exist is now ever more doubtful. Sadly, more and more, political alignment determines human relations, if not our willingness to accept objective reality.

The great irony is that over the last seventy years — my lifetime — economic growth, at least as measured by GDP, has overwhelmingly sloped upward, with only short periods of slow or negative growth. The global GDP was US$10 trillion in 1960 and $85 trillion by 2019, measured in 2015 constant dollars — a growth rate of 4 percent per year. This sixty-year period was defined by the expansion of capitalist allocation methods across the developed world, including China after 1980. That same period saw living standards rise as a direct correlation of growing energy consumption.

However, in the populations of developed economies across the Organisation for Economic Co-operation and Development (OECD), the average income of the richest 10 percent exceeds that of their poorest by about nine times, up from seven times in 1995. As a measure of inequality across countries, the U.S. average income is slightly less than $60,000 per year, while the global average income is about $15,000. In countries such as the Central African Republic, that average was less than $1,000 in 2017. This reality of income inequality has given rise to the related concepts of white privilege, reparations to historically marginalized groups, and social justice, all of which are culminating in ever more demands for redress via redistribution. Implicit in these grievances is the conviction that existing economic inequality is undeserved and socially destructive, both nationally and globally. Many contend that a higher imperative than mere financial return should constrain the right of those who hold private capital to deploy it as they see fit. In sum, the world grows ever more polarized, regardless of how much richer it has become.

In respect of climate change, we know that the most developed countries are responsible for most of the accumulated GHGs in the Earth's atmosphere. Therefore, it is argued, those developed economies should not only to transform their fundamental energy systems, but fund adaptation measures and damage claims in the nations most vulnerable to climate change effects. These economies are being challenged to reduce income inequality and simultaneously reduce hydrocarbon consumption, all without materially reducing living standards.

Developed nations face this audacious demand, to retain less of the economic output they earn and to impose new costs to deal with the climate change, while being asked to accept that leading global emitters such as China and India are exempt from any material contribution to climate change risk mitigation. These trade-offs between wealth and redistribution, between growth and fairness, and between redress and culpability have long preoccupied the various business, technocratic, and political elites that dominate the UN climate process, as well as accepted thought leaders such as the World Economic Forum, the World Bank, the International Monetary Fund, and so on; but since 2000 the specific issue of climate change has taken its place as the pre-eminent context for these considerations. This mindset is personified by Mark Carney, who demands a "financial system entirely focused on net zero." This vision subordinates the fundamental and historic purpose of finance, to deploy capital to maximize the economic interests of investors and savers. Instead of maximizing return, even if that means investing in hydrocarbons, Carney's formulation would have capital denied to that sector.

By November 2021 those who control most of the world's financial assets were sufficiently intimidated that 170 firms signed the Glasgow Financial Alliance, designed to fund investment that advances net-zero and to deny capital to investment that exacerbates GHG accumulation. This formulation is audacious. How might it be implemented? Are Larry Fink and Mark Carney to arbitrate whether capital deployments conform to decarbonization? What residual financial return, if any, will be left for the original investor?

Only one appropriate means exists to internalize climate change risk in financial or economic decisions, and that is to fix a carbon price — ideally, across the developed world. The only other option is the regulation of energy exploitation and use, and in some cases, the outright prohibition of particular energy sources. The distinction is profound and obvious: carbon prices place the onus on individuals and corporate entities to respond to a price signal, one that is, admittedly, set by government, while regulations depend solely on compliance. This route requires no price response, only a legal requirement to comply with a regulation. And, of course, zealots always prefer and advocate this option because of its brutal certainty. As experience teaches us, perverse, unintended consequences abound when imposed regulatory solutions supplant market-based ones.

If carbon pricing is the preferred policy mechanism, then what "price" is required to alter energy production and consumption patterns sufficiently to meaningfully lessen the impact of climate change? Existing economic processes used to allocate capital and optimize operations can readily internalize such a price; the cost of hydrocarbons will increase, but by an amount that reflects the consensus of democratically elected governments, and with an upper limit on the social cost of carbon. That last constraint is crucial for the resulting decisions to properly price the risk of climate change. The World Economic Forum, Larry Fink, Mark Carney, and their ilk, on the other hand, seem to imply that carbon emissions must be priced virtually at infinity — here they stand with the world's most radical environmentalists. However, to date, no developed economy has adopted a carbon pricing system reflecting such extremism, not even Canada with its proposed $170/tonne carbon tax.

The corporate sector recognizes that it must internalize the climate change risk as a fundamental determinant of corporate strategy, and corporations have been taking measures to ensure their operations conform to the requirements imposed by the relevant current and future regulatory and fiscal regime. Those regimes, and companies' responses to them, will continue to evolve. Companies can already assess fairly adequately how climate change may impact their operations physically, and whether they might need, for example, to deal directly with the risk of rising ocean levels, hurricanes, or floods on Gulf Coast refinery operations. What they cannot predict nearly as well is how the regulatory and fiscal regime will change — how radically and how punitively. This inherently political regime poses far greater risk than physical damage via climate change. Companies do not know whether they will maintain social licence to continue operations, or in other words, whether carbon taxes might increase until they can no longer produce hydrocarbons profitably.

Again, it should be the political process that determines climate policy; policy decisions should not be delegated to bankers, asset managers, or financial regulators — even if the political process has become much more polarized. And this brings us full circle: Can a world as polarized as ours see its way to applying a uniform carbon price across developed economies? The ultimate tragedy of this polarization is that there is not only a lack of common ground on the substance of what would constitute balanced climate policy, but also a lack even of a process to try to find such common ground. Are the differences in politics now too great to be overcome? Regardless, resolution to the problem of carbon pricing can only be found via the political process and the market forces that influence it.

## Failure of the Existing UN Process

Carbon pricing applied consistently across developed economies, constrained by an intellectually rigorous determination of the social cost of carbon, is advocated by virtually all credible economists as the most efficient means of dealing with climate change risk. The G7 should be the major champions of this unrealized ideal, but it has never confronted the existing UN process, much less advocated a reinvention of global climate policy based on carbon pricing.

The UN climate process has been premised from its inception on the concept of "common but differentiated responsibilities and respective capabilities," meaning essentially that countries vary in their contribution to GHG emissions, development needs, vulnerabilities, and ability to pay, a concept that sets aside the reality that GHGs have the same impact on climate regardless of where they are generated.

The logic of the Kyoto trading system, consistent in concept from that now embedded in the Paris Accord, is that some developing countries should have no, or minimal, obligations to reduce GHG emissions, even though they may be emitters. But those countries can sell their right to emit to another country that is obliged to reduce its emissions, presumably at a lower cost than what that second country could realize if it abided by its allocated emissions reduction. Of course, every emission has the same physical impact. Only in the world of UN climate treaties does the onus to reduce fall exclusively on developed countries. As noted previously, the rationale for the UN process primarily rests on an estimation of which countries bear responsibility for historical accumulations of GHGs, and which can afford emissions reductions. Although couched as emissions trading, Kyoto's system amounted to a North-South transfer scheme that sought the cheapest emissions reductions globally. The right to sell an emission would be defined by the terms of any multilateral agreement that could be struck; but an emission is an emission, regardless of where it occurs. Emissions reductions simply are not real unless combustion does not occur, or unless emissions from combustion are sequestered. The same logic applies to carbon sinks: allowing a country to sell its willingness not to deforest is only possible if that country has been granted permission to deforest in the first place.

Again, the allocation of reductions targets and permissions to sell reductions was driven as much by the objective of North-South transfer as by the relative cost abatement curves of respective countries: that is, the relationship between levels of emissions reductions and their cost. But this system broke down as early as Kyoto, when China and India were exempted from any emissions reduction obligations despite being amongst the world's top five GHG emitters, a matter at the heart of the U.S. Senate's vote not to ratify the treaty. This free rider problem has bedevilled the UN process ever since, even at Glasgow. Failing to impose targets on two of the world's largest three emitters of GHG, regardless of their claims for exemption on the basis of history and self-interest, dooms the current UN process to failure.

Between the conferences in Kyoto and Paris, global emissions grew on the order of 40 percent, as measured by the NOAA's Annual Greenhouse Gas Index, and this increase occurred despite emissions levels stabilizing substantially across the developed world. China is now accountable for almost 30 percent of annual global emissions. The UN climate process, so far, has largely failed to meet its basic objectives, if measured only by the level of global emissions. In a more perfect world, one typified by internalizing a uniform carbon price constrained by the social cost of carbon, could the results have been any worse?

At Copenhagen's 2009 climate conference, the process nearly broke down completely. Instead of developing a new treaty to replace Kyoto, the meeting led to increased tensions between developed and developing countries surrounding financial assistance for adaptation and mitigation. Both Canada and the United States essentially restated their Kyoto commitments for emissions reductions, but the pledges remained non-binding and, in Canada's case, difficult if not implausible to achieve. Of course, the UN process continued after Copenhagen, and for many, the process of dealing with climate change was clearly as important as the substance of dealing with it reasonably.

Between Copenhagen and Paris's 2015 meeting, the Obama administration committed itself to achieving a result that would not carry the same stigma of failure. As mentioned earlier, the resulting Paris Accord was designed to circumvent the U.S. Senate — the accord was an agreement, not a treaty per se. Again, China still did not make any real reduction pledge. U.S. negotiators decided not to press the point and allowed each country at the table to pledge reduction targets of its own devising. As for a North-South transfer, the agreement mentioned that developed nations would provide funding of US$100 billion to accelerate adaptation in vulnerable countries, Of course, global, coordinated carbon taxes could have provided some of the funding. But that was not to be.

Paris was hailed as a great success, mostly by those with a stake in the UN process perpetuating itself. The signatories had accepted temperature containment as their fundamental objective — specifically, to "substantially reduce global greenhouse gas emissions in an effort to limit the global temperature increase in this century to 2 degrees Celsius above preindustrial levels, while pursuing the means to limit the increase to 1.5 degrees." Although the actual pledges and targets were insufficient to meet that temperature containment, Paris created a commitment to ever more process. Still, the free rider problem went unaddressed. When the parties met again in Glasgow in 2021, these fundamental flaws persisted, but the aim had shifted from containing 2°C to containing 1.5°C — de facto decarbonization.

## Semantics and Narratives

The language applied to the climate change risk grows ever more reckless, while media and politicos adopt ever more hysterical narratives, ignoring more reasoned, dispassionate considerations of cost and benefits. Consider these examples of terms so prevalent that they have become practically ubiquitous in conversations about climate change.

First: *existential*, as in existential crisis, or existential risk. Nowhere in the roughly four thousand pages of the Working Group 1's contributions to Assessment Report 6 does this term occur. Moreover, the report actually ascribes a low probability to any event occurring before the end of this century that would trigger warming at a rate unaddressable by mitigation or adaptation. Yet, much of the media gives credence and airtime to commentators who are neither climate scientists, nor formally trained in risk assessment, nor have any background in the energy industry, and who speak flippantly of the "existential" crisis requiring immediate collectivist intervention.

Second, consider the word *catastrophe*. The IPCAA AR6 offers no specific quantified specific definition of that word applied nationally, let alone globally. It uses the terms *catastrophic* or *catastrophe* five times, all qualitatively. Must a climate event that produces irreparable property loss be considered catastrophic? Or must recovery cost a minimum amount? Or does a catastrophe by definition involve loss of life or some attributable and irreparable biodiversity impact? Neither the IPCC nor the media has set any objective quantitative standard.

Media and political leaders glibly apply the term *crisis* to describe the status of climate change risk. If they mean that we are close to a tipping point, then I concur with their use of the term. I must stress, however, that I do not believe that we have reached a climactic tipping point, one that would lead to the Earth becoming uninhabitable; rather, I believe that policy responses to climate change risk across the developed world are about to tip over into a destructive force that may profoundly reduce global living standards. The impetus to impose decarbonization has placed us at the brink.

Those who contend that a physical crisis is imminent do so without validation from the IPCC reports. To quote again the ARC Physical Science Report: "Proportionate to the rate of recent temperature change, there is no evidence of such non-linear responses at the global scale in climate projections for the next century, which indicate a near-linear dependence of global temperature on cumulative GHG emissions," while, "for global climate indicators, evidence for abrupt change is limited." My takeaway is that we have time to replace decarbonization with more optimal policy on mitigating the climate change risk.

All of this hyperbolic language rests on the contention that "tail risk" — again, the risk of events with low probability but high impact — relating to global climate change is too severe to ignore. The mere possibility, no matter how unlikely, of extreme, irreversible climate change demands, by this argument, decarbonization, regardless of the cost to human welfare or economies. The AR6 reports that some extreme weather events have increased in frequency and intensity due to human-caused emissions of GHGs. However, it does not follow that when a specific extreme weather event occurs it can necessarily be attributed to climate change. Nor do we know how, over time, adaptation investments will moderate those impacts.

Another point of genuine disinformation is the assertion that decarbonization comes with no real cost either short- or long-term, or at worst modest and acceptable costs. Believers contend that energy transition offers only upsides: new investment opportunities and new jobs, all emitting fewer GHGs. The Trudeau regime is eager to resort to this kind of invocation. But, truth be told, this transition means essentially destroying existing energy systems predicated on hydrocarbons, from production to end-use infrastructure — systems that work and have worked affordably, and may continue to do so even if they have to absorb a carbon tax that reflects increasing climate change risk.

Furthermore, this narrative invokes a transition to available, affordable, and entirely functional renewables and batteries, seldom acknowledging that renewables have yet to overcome the fundamental issue of intermittency on any credible net-zero basis. Renewables are not always available, full stop. The technology required to store surplus electricity from renewables when they are producing is limited. Renewables need backup, typically from natural gas, which of course emits carbon. So, unless natural gas backup is part of some affordable CCS system, which is not yet the case, intermittency cannot be dealt with in a manner consistent with net-zero or, more honestly, decarbonization. Natural gas remains indispensable to ensuring that renewables-powered energy systems remain reliable, the dreams of California Democrats notwithstanding. Trying to ignore the real constraint of intermittency simply amounts to dishonest sophistry.

Grudgingly, in early 2022 the EU designated natural gas and nuclear as "sustainable" energy supply options, recognizing implicitly that renewables alone cannot support an entire electric generation system. Regardless of how inexpensive wind or solar may be when they are available, their cost is infinite when they are not. Some stored electricity must be able to be sourced in those periods — via batteries, or more likely, systems must resort to natural gas generation.

We also hear little about the massive transmission infrastructure required to bring remote renewable power to the point of consumption — not in the form of buried pipelines, but as above-ground transmission lines. As more end uses rely on electricity rather than fossil fuels, and if alternatives such as natural gas and nuclear are phased out, electric grids will grow less reliable. Also, the massive expansion of non-emissive electric generation, transmission, and storage will not avoid the classic obstruction of NIMBYism, which has so devilled hydrocarbon infrastructure. Various instances of opposition to the construction of such infrastructure have occurred across North America in recent years, mostly notably in Maine, where, as noted earlier, residents vetoed the construction of an electric transmission cable project to bring Quebec hydro power to New York City. As well, a world based on electric storage requires certain key minerals, and these will be subject to the same kinds of constraints and objections as hydrocarbons. The public needs real information, offered with intellectual honesty, but our media and politicians offer none.

We already know that available technologies offer no perfect substitute for hydrocarbons. Renewable technologies use minerals including lithium, copper, and cobalt, and the required mining projects will face resistance over environmental degradation, collateral health impacts, and labour practices. Moreover, the IEA confirmed in May 2021 that China has acquired a majority share in the global processing of cobalt, lithium, and the rare earth elements needed to make electric vehicles. Increasing silicon supply from northwest China, required to make solar panels, relies on a coerced workforce of Uyghurs, numbering in the hundreds of thousands. Something else to be considered is the land mass that will have to be redeployed to install expanses of solar panels and wind turbines. Populations will resist, and species will be impacted.

Furthermore, the majority of world food production relies on fertilizers made of the chemical processing of natural gas. Most fertilizers now fundamental to sustaining world food supply are based on ammonia, a chemical made up of nitrogen and hydrogen. Accessing hydrogen economically requires reforming natural gas — splitting methane into hydrogen and carbon dioxide — so, global ammonia manufacture is a significant GHG emitter. Systems based on the electrolytic separation of water, "green hydrogen," are massively expensive relative to conventional natural gas reforming. Of course, the same issue affects whether CCS systems can be deployed at the back end of each reformer to keep carbon dioxide out of the atmosphere. Again, at what carbon price does that process become feasible? And can the world afford the increased cost of its food supply?

As well, individuals will be constrained in how much energy we can consume and in what form, regardless of whether we have the means to afford such consumption. We will no longer be free to drive private vehicles, to eat what we like, or to decide for ourselves whether to retrofit our homes based on what we can afford and how we wish to spend our money. All this is the logical consequence of the mentality that avoiding any GHG has infinite value. The *reductio ad absurdum* of decarbonization becomes obvious — no personal carbon emissions are too trivial to eliminate. Will there be any limit on the intervention into our lives? Has the media ever really put this question to the glib advocates of decarbonization?

## Security and Resistance via Hydrocarbons

Russia's invasion of Ukraine in February 2022 surely made clear the fundamental security risk represented by giving a regime like Putin's control of so much of the energy supply, particularly natural gas, to much of Europe. The logical consequence of Russia's actions should be for North America to supply Europe with natural gas instead, though that transition will take time. For that to happen, Canada and the United States must value security considerations when they process approvals of the entire value chain of investments required for production, transmission, and liquefaction of natural gas. Instead of rejecting future developments as inconsistent with decarbonization, can North America genuinely reprioritize international security and resist totalitarianism? Expanded North American LNG and crude oil exports must still be subject to climate policy, but can pressing geopolitical matters become central to reconsidering the extremism of decarbonization?

## The Ideal Is the Only Answer

The current political class of the developed world is committed to decarbonization, but only the most affluent countries of North America and Western Europe will likely even attempt to seriously effect it to some degree. Others, such as China, will likely do nothing substantive to alter their energy consumption, in terms of either continued hydrocarbon use or absolute levels of consumption. A real dystopia looms, in which real democracies are economically diminished, while autocratic free riders enjoy that sacrifice. The reset we need should fundamentally follow the policy outline offered by William Nordhaus. Here, I outline the key elements of his article in *Foreign Affairs*, "The Climate Club," along with some of my own elaborations.

First, the UN climate process gives way to an agreement within the G20, including China and India, to impose the same level of carbon tax on GHG emissions generated within their own countries. The G20 members unanimously agree on the price level contained in the tax, but it is constrained at its upper limit by a collectively accepted social cost of carbon, determined in turn by a mechanism prepared by the G20 countries themselves. The price begins at US$50 to $100/tonne and is redetermined every five years. This carbon tax comprises the pre-eminent carbon policy instrument, and individual countries can do more, but are not compelled to do so. Outside the G20, countries face import/export adjustments if they do not impose the carbon tax. If one or more of the G20 countries do not agree with a specific level of tax, then the remaining countries that do agree may proceed without them, and once they agree to the terms, the G20 countries have a year to implement the tax. If no agreement can be reached then, however, the entire process breaks down. A portion of the tax is allocated to deal with adaptation, damage, and loss in nations with legitimate claims for such funds. The process the UN uses now, of preparing technical reports to assess the status of the physical science, and furthermore to assess the economic and social impacts of various carbon-pricing scenarios, continues. Finally, the G20 accepts that any climate intervention more costly than the social cost of carbon must not be imposed. This will prove a real constraint on the resort to regulations and interventions that implicitly incorporate a social cost of carbon in excess of the governing determination.

This formulation would deal with the free rider issue; otherwise, the option for a Mexican stand-off would persist. Just as importantly, the ratification by each country would ensure that the actual level of carbon tax had democratic sanction, at least by the democracies that dominate the G20, and would require that existing fiscal arrangements must accommodate this tax. Again, myriad economic incentives would arise for energy efficiency and research spending in the context of a defined, objective price to emit carbon. Transparency and uniformity remain inviolate design features. The design would further ensure real political validation for dealing with climate change, because the voters could determine if they were prepared the impose the transparent carbon tax on themselves. Bureaucracies would be discouraged from imposing regulations and interventions not otherwise justifiable within the constraint of the social cost of carbon, and this would apply as well to banks, corporate boards, and regulatory entities especially in the financial sector.

Winston Churchill once said that "perfection is the enemy of progress." Pragmatism and incrementalism should win out, in other words, over idealism and ideological purity. This is usually sound advice, but sadly, in respect of climate change, we have now reached a point where resorting to the "perfection" of carbon pricing uniformly applied across developed economies constrained by the social cost of carbon is the only rational alternative to avoid the real harm decarbonization will otherwise wreak on world economies.

# CHAPTER 9

# Oh, Canada!

Early in the proceedings of the 2021 Glasgow climate summit, Prime Minister Trudeau chaired a session on the concept of global carbon pricing using carbon taxes. Commendably, he advocated for "a clearer call to create a global standard around putting a price on pollution. Not only will that encourage innovation," he stated, "it will give that clear price signal to the private sector that making the right capital investments to transform to lower emissions makes sense. It also ensures that those who are leading on pricing pollution don't get unfairly penalized." This statement is entirely valid, alluding to Canada's vulnerability with a pricing stringency too high and unmatched. Trudeau was acknowledging the legitimate concern of maintaining Canadian economic competitiveness. However, he made no link to the social cost of carbon, nor to how other climate intervention and regulation could be rationalized to such a global carbon tax. At no time did he call for reconsidering the UN process, nor did he question the objective of decarbonization itself. At best, his session afforded a unique opportunity to exhort, implicitly, Canada's major trading partners to consider matching the country's $170/tonne carbon price planned for 2030.

Of course, Canada's ever-zealous, extremist environment minister, Steven Guilbeault, boldly asserted his intention to implement a massive system of border tariffs on imported goods from major trading partners that fail to impose a comparable carbon tax stringency — a move that ignores the obvious question of why Canada would impose such an extreme cost on itself in the first place when those trading partners cannot. This is especially baffling when the United States cannot even manage a national carbon tax at all, let alone one of the stringency that Canada has proposed.

One wonders if Trudeau has really seriously considered the potential for reinventing the entire UN process around a global carbon tax conditioned by the social cost of carbon — or was the Glasgow panel event merely a polite, nudging gesture implying no real commitment to the concept of global coordinated carbon pricing? This incident so classically exemplified Canada's position on climate change and on the UN process related to it — this country seemingly has no capacity to recognize its own economic self-interest. It seems infinitely willing to abide climate free riders and embraces national emissions reduction targets that require deconstructing its hydrocarbon production industry in a world that shows no capacity to reduce hydrocarbon demand. Canada has national emissions reduction targets that require carbon prices more extreme than any other developed economy has come close to accepting. Is this altruistic sacrifice, or just stupidity? And how can a Canadian electorate tolerate this?

One might reasonably hypothesize that roughly a third of Liberal voters in the most recent federal election in 2021 were affluent enough to share the basic economic interests of the 40 percent of the Canadian electorate that consistently voted for right-wing alternatives over the last three federal election cycles — alternatives that span the gamut from outright climate change denialism to policy incoherence to a minimalist approach to intervention via carbon pricing. The federal Conservatives of 2021 might be described, generously, as carbon policy minimalists. But today's Liberal Party is not that of Robert Winters, let alone Jean Chrétien; it has become a centre-left party, with only a residual minority of supporters who might consider voting for a federal conservative alternative, even the 2021 version. One can only deduce that those potential conservative voters have not, as yet, discerned any serious economic risk from climate policy to Canada or to their investment portfolios. Their failure to see this becomes ever more inexplicable, given that throughout the 2021 federal election the Liberals espoused decarbonization, which means destroying the 5 to 10 percent of the current economy attributable to hydrocarbons. As a seasoned poker player once remarked: lessons will cost extra. So they will, for Canada.

The Trudeau government's position on climate and hydrocarbons can be traced back to 2009 and 2012, when Gerald Butts laboured in the Canadian ENGO community, trying to resist Canadian hydrocarbon development and growing increasingly resentful toward not only that industry but also Alberta, the most American of Canadian provinces and the most defiant of leftist dogma, Medicare notwithstanding. Butts and Trudeau ascended to power in late 2015. This victory was not, of course, based solely, and perhaps not even mainly, on the climate and energy policies the Liberals espoused; they also benefited from, among other things, the NDP's inexplicable collapse, disenchantment with Stephen Harper after almost a decade, and efficient concentration of their party's vote. But their espoused policies and promised actions relating to energy and climate were well aligned with the long-standing agenda of Canadian ENGOs — basically to frustrate Canada's hydrocarbon production potential while refusing to concede that massive economic value destruction must follow. All this I have discussed at length in my previous two books. The list of blows to Canada's hydrocarbon industry is long and tragic, and I laid out the most recent examples in part one of this book:

- The Oil Tanker Moratorium Act, which bans oil tankers off the B.C. coast north of Burrard Inlet and effectively eliminates hope of restoring the Northern Gateway project
- Bill C-69, which changes how major infrastructure projects are reviewed and approved in Canada, essentially making the regulatory approval risk impossible for private capital
- Imposition of the oil sands emissions cap, the only economic sector in Canada facing such a constraint on its own expansion
- Denial of Northern Gateway after it gained regulatory approval following almost eight years of due process

- Regulatory treatment of Energy East and Petronas LNG, leading to their cancellation despite fundamental economic commercial support
- Imposition of the Canada Fuel Standard, requiring the use of "bio-fuels" in the Canadian gasoline pool, at costs exceeding even the 2030 national carbon price of $170/tonne
- Abject acquiescence to the denial of KXL by hostile Democratic regimes
- Acquiescence to fracking bans in Quebec and New Brunswick
- Unconditional acceptance of net-zero emissions

The Trudeau government's 2015 compromise with the Alberta Notley and B.C. Horgan governments on TMX and LNG Canada, respectively, have become ever more difficult to sustain with Guilbeault in Cabinet as environment minister and former renewables developer, former NDP bureaucrat, and previous environment minister Wilkinson as minister of natural resources. Guilbeault is implacably committed to the deconstruction of the Canadian hydrocarbon industry, with no engineering, economic, or scientific credentials, while Wilkinson is similarly committed to deconstructing the very industry whose interests he is meant to represent in the deliberations of the Canadian Cabinet and has no relevant experience in the energy industry. These two ministers of the Crown, elected to serve all Canadian regions and interests, seem bent on trying to undermine an industry that supports thousands of Canadian workers and provides substantial revenue to governments, federal and provincial. They do so while Canada sits on world-class hydrocarbon production potential, in its oil sands and western shale gas reserves. The country is failing to realize tangible economic value in a world that shows, as yet, no capacity for diminishing its demand for hydrocarbons, Glasgow notwithstanding.

But has the Trudeau government ever acknowledged the value Canada has lost or indicated how much more this country will have to bear if we implement decarbonization? Why would it? Of course, instead, we hear more unjustified claims that value will arise from the energy transition ahead, and we watch funds directed to the promoters of CCS, renewables, biofuels, and retrofit industries, who clamour for explicit subsidies and grants instead of simply financing their low-carbon "good works" by the value of avoided carbon tax payments, especially at the $170/tonne level. Indeed, the Trudeau government has no reason to be intellectually honest when it can instead glibly demonize the hydrocarbon production industry, and thereby demonize Alberta implicitly as well. Most of the burden of reduced emissions, in the short and medium term, will occur by reducing exports, or by imposing regulations that render existing production uneconomic, such as extreme methane containment limits, CCS for any steam production related to in situ oil and gas production, and any existing natural gas reformation for either refinery upgrading or fertilizer production. And yet, the world will continue to demand as much oil and gas as ever, regardless of Canada's economic self-destruction. The situation looks even worse when we consider that Canadian LNG and heavy oil production create fewer emissions than most of their global competitors — competitors that will pick up Canada's market share.

○ ○ ○

Nothing is as grim or more eloquent than the letter written by Guilbeault and Wilkinson to the leaders of the Net-Zero Advisory Body (an entity with no credible industry representation) about their intentions to deconstruct Alberta's hydrocarbon industry:

> A clear area of concern is emissions from the oil and gas sector. Greenhouse gas emissions from the sector have risen 20 percent since 2005 and now make up 26 percent of Canada's total emissions, making oil and gas the largest emitting sector in the country. Getting to net zero starts with ensuring that emissions from the sector do not increase. Canada's commitment to cap and cut oil and gas emissions is a first in the world for a major energy producer.

> As you may know, we made a commitment to Canadians to ensure that the oil and gas sector would reduce emissions at a pace and scale needed to achieve net zero by 2050, with five-year targets to stay on track to achieving this goal. To help us get there, we also committed to set 2025 and 2030 milestones, supported by the advice of the Net-Zero Advisory Body. This will help to ensure that reduction levels are ambitious and achievable and that the oil and gas sector makes a meaningful contribution to meeting the nation's climate goals for 2030 and 2050, while taking a people-centred approach to ensure a just transition for workers and their communities. Canadians gave us a clear mandate to deliver.

No other economic sector or region of the country was so specifically singled out. Despite the fact that oil and gas emissions from actual oil and gas production represent less than 20 percent of national emissions, the industry is expected to deliver the vast majority of emissions reductions to meet current national targets. A diminished economic contribution from Alberta impacts all of Canada, and no one has offered a credible answer to how the historic contribution from hydrocarbon production will be replicated. It is always convenient to demonize that segment of the country with the least political value; the federal government is compelled to imply that its electoral base is blameless and that others should pay. For Guilbeault, this is not subtext, but a core belief and statement of intent.

Alberta, meanwhile, will suffer further assault over how it generates its electricity and over its agricultural sector. Guilbeault will surely demand that Alberta cease using any hydrocarbons, whether coal or natural gas, for electric generation; it is only a matter of when, and my guess is mid-2022. How, exactly, would that capacity be replaced — with more renewables and batteries? The wind hardly blows at -30°C, nor does the sun shine for more than eight hours a day in the depths of the Alberta winter. Nor do heat pumps work well at those temperatures; there is simply less heat to transfer out of the atmosphere, regardless of how much electricity we consume.

Importing massive amounts of hydro power from British Columbia or Manitoba could only work if those provinces had undeveloped capacity on the necessary scale, and if the incremental dams and transmission required could gain regulatory approval by the end of the decade. Bill C-69 would bedevil its Liberal proponents, in a dose of richly deserved karma that offers only cold comfort to observers. Would the province simply go nuclear? But the same issue of regulatory approval would apply. How many plants would Alberta need to ensure reliable electricity? If renewable increased the overall cost, would they still be part of the system capacity? Alberta would be expected to abide increased cost and less reliable energy than natural gas offers, even while absorbing applicable carbon taxes. But then, according to the mindset of Guilbeault et al., no emissions reduction is too costly to impose — especially in Alberta.

In late March 2022, Guilbeault officially announced that Canada would commit itself to a 42 percent emission reduction target by 2030, imposed on the hydrocarbon production sector — fait accompli, notwithstanding the net cost to Canada, or especially to western Canada, the impact on Canadian competitiveness or the geopolitical implications. In his own words: "There is clearly a crisis in Ukraine, just like we're slowly emerging from the Covid crisis, and there will be other crises in the coming months and years. But climate change will not go away, and if we're thinking we can solve the crisis by exacerbating another one, those people who think that are clearly mistaken."

Brutal and unconditional.

But further interventions loom with this mindset. Alberta also has a significant red meat production industry, representing about 20 percent of agricultural production. No doubt Guilbeault will want it shut down by the end of the 2020s. Vegetarianism is the perfect adjunct to radical climate policy. Indeed, looking ahead from mid-2022, Alberta's prospects look bleak, its fundamentally world-class industry soon to be diminished or outright destroyed, while the costs of utilities massively increase, all in the name of meeting Canada's net-zero emissions objective (nay, obsession).

In fact, this situation looks bleak not only for Alberta, but for Canada as a whole. Yet who could credibly deny that if these hydrocarbons were in

Quebec, rather than in western Canada, the country's climate policy would have taken a very different course over the last three decades? Canada would never have aligned itself so altruistically and completely with the UN process, but would probably have behaved more like Australia has, participating only grudgingly. Moreover, Canada would have exempted more resource rents from its equalization system, instead of forcing an eastern province to share its hydrocarbon wealth as it has forced Alberta. Such is the cruel reality of Canadian federalism and demography, made even worse by the twisted antipathies of the likes of Guilbeault and the unthinking acquiescence of the likes of Trudeau and Freeland.

Sadly, the Canadian electoral arithmetic seems unlikely to change, with a Liberal coalition in the Atlantic provinces, the Montreal area, and most of the rest of urban Canada. These votes may amount to only 33 percent of the popular vote, but they are concentrated in such a way that they translate into many seats in the Canadian Parliament. That 33 percent has, over Trudeau's tenure, guaranteed him a majority or de facto majority government, which means his party faces no constraints, since Canada has no entity comparable to the U.S. Senate and never will. And so, Alberta will carry most of the economic cost of Canada's efforts to reach decarbonization — a reality with no political consequences for the Liberals. Some will say that Alberta's dreary fate is merely a consequence of geology and nothing personal — as though seventy years of alienation from the federal Liberals is just an irrelevant coincidence.

○ ○ ○

Decarbonization offers no net upside for Canada. Ironically, Canada stands to benefit significantly in a 3°C world, with longer, warmer growing seasons, less expensive winterizing for businesses and households, and continuing hydrocarbon production. Not once has a Trudeau government told Canadians the cost of meeting the country's emissions reduction targets or how that compares with other countries, most significantly the United States. Trudeau is unlikely ever to tell his populace the truth, that Canada made all possible "cheap" emissions reductions to electric generation infrastructure long before countries received any credit for such decisions.

If Canada were guided more by economic self-interest, or even led by more economically literate politicians, it would have long espoused the following basic principles for climate policy:

- Canadian climate policy must be defined solely by carbon taxes.
- Those taxes must grow no higher than those Canada's major trading partners impose on themselves, explicitly or implicitly.
- No other market intervention may be used to accelerate "low carbon," full stop. No mandates, no fuel stands, no subsidies, no regulations.
- No explicit emissions reduction targets may be imposed, either nationally or sectorally.
- Some of the carbon tax proceeds must be reserved for adaptation spending domestically, and eventually internationally.
- No household rebates can be given, because these only blunt the whole point of the tax and obscure the political legitimacy of dealing with climate change.
- Canada should actively engage to fundamentally change the UN climate process to one based on carbon pricing via carbon taxes rather than mandated emissions reductions across developed countries.
- An appropriate system of import/export adjustments must be implemented if Canada must sustain trade with countries not applying the same carbon tax stringency.

I would take no issue with Canada challenging its major trading partners to implement $170/tonne as the going tax; that level represents the equivalent of crude oil prices near $70/bbl, about the price of oil by the end of 2021. If those countries could not relate to, let along implement, this stringency, that should prove instructive to Canada as the country decides whether to persist with imposing the cost on itself. As for weather events, Canada should approach these as an adaptation exercise, whether in the form of improved forest management practices, better civil engineering when constructing roads and dikes in areas vulnerable to flooding and landslides, or even better winterization practices. The funding would, of course, ideally come from the national carbon tax. For context, exporting an additional one million barrels per day of Canadian oil, with a netback to Alberta of US$60/bbl, would represent almost $30 billion of revenue annually. If the pricing level were sustained over time, that amount could be realized, and even 5 percent of that would provide significant funds for dealing with domestic climate change–related adaptation and remediation.

The fundamental question for Trudeau and Guilbeault remains: Is any cost too high to reduce GHG emissions in Canada? Does any other consideration matter? Do any conditions apply? If this book has any utility, it will be to evoke answers to these questions — answers that Canadians deserve. But perhaps we can anticipate the answers. In late November 2021, Dr. David Suzuki reacted to the failures of Paris and Glasgow, and to the temerity of Canadian climate policy, ominously musing, "We're in deep, deep doo-doo. And the leading experts have been telling us for over forty years. This is what we've come to. The next stage after this, there are going to be pipelines blown up if our leaders don't pay attention to what's going on." When commentators read his comment as a threat, however, he clarified: "The remarks I made were poorly chosen, and I should not have said them. Any suggestion that violence is inevitable is wrong and will not lead us to a desperately needed solution to the climate crisis. My words were spoken out of extreme frustration, and I apologize." Nevertheless, his empathy with the animus at the core of the environmental movement was clear. Instructively, and incredibly, neither Trudeau nor Guilbeault condemned Suzuki's musings.

Sadly, the Canadian right has behaved just as dismally, never articulating an intellectually rigorous counter-argument to decarbonization. For too long, the right resisted carbon taxes, when for Canada they offered the only policy option that could ensure international competitiveness. Nor have Canadian conservatives clarified that this country's carbon pricing must not significantly misalign with the United States'. Instead, when it comes to climate change, the right has mired itself in incoherence, trying to pander to leftist voters whom they will never win over, while refusing to completely commit to carbon pricing.

How much value must Canada lose before some political realignment may occur? Unlike some other countries, Canada does not face a physical shortage of hydrocarbons over the short run, the Line 5 idiocy of Michigan Democratic governor Gretchen Whitmer notwithstanding — she has threatened to close down the major crude oil and refined products pipeline into Ontario despite the fact that the pipeline's operation is a matter of federal jurisdiction and that it has long-standing approval to operate. Other countries have learned, over the winter of 2021–22, what energy transition really costs when hydrocarbon supply is truncated, but in Canada the effects will take longer to manifest. When they do, however, those effects will affect Canada's economy even more. How can a country replace 10 percent of its economy and hope to sustain the same economic contribution? What unknown source of competitive advantage is supposed to kick in? How can a country that already derives over 50 percent of its GDP from the public sector dismiss one of its few internationally competitive private-sector industries? So glibly, so fatuously? So ignorantly? So spitefully?

If the fundamental merits of a more rational cost/benefit–based climate policy and economic self-interests are not compelling enough, then Canada should be convinced by the emerging geopolitical imperative of increasing hydrocarbon production domestically in order to reduce the leverage gained by countries that oppose the world order we value — and that we should value more than decarbonization. The threat posed by Russia beginning in February 2022 makes hydrocarbon production Canada's greatest imperative since the Second World War — even a moral imperative. This requires a fundamental change to the Canadian left's mindset toward domestic energy development, which is typified by obstruction as its fundamental intent and process. The sad consequences of this I laid out in chapter 4. Some of the key losses for Canada were Northern Gateway, Energy East, Petronas LNG, and, most notably, KXL. The dynamic also manifests itself in respect of the protracted construction processes of Coastal Gas Link and TMX.

# CHAPTER 10

# Hope and Realism

The world urgently needs to reconsider decarbonization as the global objective to deal with the climate change risk. As events continue to unfold, I become only more passionate about, and committed to, that objective, if that is possible. Accepting that we must adapt to some potential climate change typified by a future global temperature increase approaching 3°C is the optimal objective, if the world's economies, and in turn global human welfare, are not to be diminished materially. More to the point, the world is not at a point that necessitates decarbonizing, notwithstanding the pledges made at Glasgow, and notwithstanding the continuing aspirations of the global environmental movement or the blandishments of feckless political leaders such as Justin Trudeau, Joe Biden, and Boris Johnson.

The likes of China, India, Brazil, Mexico, and Russia are not going to decarbonize to any material degree, regardless of what North America, Britain, or the EU countries do. Emissions in these still developing countries are rising and will continue to do so for the foreseeable future. Those countries are not about to sacrifice economic growth or accept decarbonization that would make the energy required to sustain that growth unaffordable and unreliable. Whether the G7 countries attempt decarbonization in the short run will not matter, as emissions will continue to rise regardless, and eventually the world will reach a point at which adaptation to 3°C temperature increase is the only recourse. Any credible forecast post-Glasgow validates that looming reality. In the interim, how much cost will be imposed on developed economies in the name of decarbonization? How much will we destroy energy systems that work and replace them with less affordable, less reliable substitutes? What degree of venal, self-interested exhortation will we tolerate from those who stand to gain from the trillions about to be spent on trying to achieve decarbonization? Can this value destruction at least be minimized? Better yet, can we accept a cost of climate mitigation and adaptation that is more optimally matched to the actual risk itself?

What might offer some optimism that this momentum to decarbonize can be mitigated, if not stopped? First, markets assert themselves, especially energy markets. The cost of the energy transition required to effect decarbonization cannot be hidden from consumers or taxpayers, and the costs are substantial and immediate, exceeding even Canada's $170/tonne. During the fall and winter of 2021 and 2022, in Europe and China the price of hydrocarbons escalated due to past constraints on supply and related infrastructure. Relying on hydrocarbons, even coal, proved inescapable to keep basic energy demands kept. Even Germany, committed as it was to energy transition, had no recourse but to rely on natural gas, even if that meant greater geopolitical reliance on Russia. This was all a testament to the reality that, due to their intermittency, using renewables without resort to natural gas is practically impossible. The energy density and ease of storage of hydrocarbons offers an efficiency that cannot be replicated economically. The full-cycle carbon balance of renewables is, moreover, more problematic than most observers appreciate; the energy required to produce the components of solar and wind systems cannot be evaded. If decarbonization is the actual goal, the attendant costs are enormous, whether measured in terms of affordability or reliability, or in terms of abandoning current living standards. Those costs cannot be hidden, and political reaction is inevitable.

Second, democracies work provided electoral processes are fair and insulated from fraud or intimidation. As the cost of energy increases over the short and medium term due to efforts to effect decarbonization, electorates will react negatively, implicitly questioning how much risk mitigation they are expected to bear — or bluntly, questioning if there is any mandate for decarbonization at all. Such an extreme alternative should proceed only on the basis of explicit democratic sanction, ideally via direct referendum. A rational response to such referenda will emerge once the costs of trying to decarbonize become explicit and the impacts on the world's developed economies become more widely known, relative to the benefit of climate change mitigation that is largely in the future and in locations separate from those that must bear most of the costs.

The constraints of the U.S. political process continue to impede even the current intention to spend up to $500 billion in tax credits for renewables, electrification in vehicles, retrofits, biomass fuels, and transmission; it is far from certain that a bill to that end will even be passed, let alone a bill that mandates transforming the country's electric generation sector by 2035. Senator Joe Manchin's explicit rejection of those provisions, along with complete Republican resistance at the end of 2021, likely confirms the rejection of the Biden agenda to deal with the climate change risk. If the United States cannot commit itself credibly to decarbonization, who else can? Despite the American left wanting so ardently to use climate as a basis of social transformation, the reality is that, even today, without the full cost of decarbonization yet asserting itself, the country is split evenly for and against decarbonization. Unless the mechanics of the U.S. Senate are changed, even a fifty-fifty vote represents a veto on decarbonization at the federal level. How can Canada be committed to decarbonization if the United States is not?

It is not up to the financial community and financial regulators to take on the climate agenda by denying capital or forcing divestment, or by forcing extreme operational costs on hydrocarbon production. In a democracy, governments determine on what terms an industry should operate, and that role should not be delegated. Again, governments should price GHG emissions and allow markets to internalize that price. My hope is that financial institutions will rediscover what their actual roles are and resist trying to impose a specific agenda that can only be validated by the political process and implemented by government.

As well, many elements of what is known as "mainstream media" approach the climate change issue with more balance and regard for economic realities, instead of seemingly robotic advocacy of decarbonization. Unaffordable and unreliable energy costs are real and immediate. They deserve balanced attention as much as climate risk does. For emphasis, the free riders, notably China and India, have clarified they will not embrace decarbonization over the next thirty years, rejecting the trade-off of diminished economic growth for decarbonization. Glasgow, to any realistic observer, made this obvious. The world will simply have to adapt to 3°C, but hopefully in the context of uniform carbon pricing conditioned appropriately via the social cost of carbon. How can the existing UN process persist?

One can only hope that a majority of the electorates in the developed world come to recognize that the environmental movement, especially in respect of climate change, is no inviolate entity beyond challenging; indeed, it should be questioned closely, in terms of both its motives and its methods. The most extreme application of decarbonization would see the world become more primitive, isolated, and static. Decarbonization represents a denial of humanity's capacity to adapt and innovate, and not merely to sustain existing living standards, but to improve them materially over time. At the heart of the environmental movement may be a fundamental anti-human animus, integral to the notion that capitalism is unsustainable, despite what human history tells us, and that the consumption that drives it is unjustified if not outright immoral. Thus, the movement envisions a bleaker and more dispiriting world, all made to come about to deal with the tail risk of climate change. I can only hope more people come to appreciate this.

○ ○ ○

How much value destruction will have to be endured before a more optimal objective than decarbonization emerges in the developed economies of the world? The most economically efficient policy can also be the more politically sustainable one: a policy based on uniform carbon pricing via carbon taxes across the developed economies of the world, with an associated system of import/export tariff adjustments to account for trade with countries that have not imposed the same tax on their national emissions. A tax constrained by a realistic expected value assessment of the social cost of carbon. Moreover, the revenues from these carbon taxes can fund adaptation, both domestically and internationally. Adaptation takes on many forms to deal with more frequent extreme weather events, from improved forest management practices to various civil engineering innovations, to facilitating migration to less vulnerable locations.

This reinvention based on carbon pricing must apply to all emissions equally; only then can we dispense with subsidies and regulations for low carbon. If low-carbon technology of any kind becomes economic simply by internalizing the carbon price, so be it, and if hydrocarbons can internalize the same tax and still maintain market share in global energy demand, so be it as well. In this formulation, financial industries and regulators need not overstep their traditional role by trying to become de facto agents for climate policy. All economic players would react to the same price and the same price horizon, and the same upper limit on price. Capital allocation could then unfold in the traditional manner, which has served economic growth and efficiency, as demonstrated by the record of the last fifty years.

As I have asserted, Canada could become a constructive force to bring about a reinvention of the entire UN climate process; however, Canada will more likely have to assert its influence on the G7, asserting that this entity should be the nucleus of change in dealing with climate change risk, based on adaptation and carbon pricing. It is understood that the nations of the G7 have been the major historical emitters, so a special moral obligation rests on them to find optimal climate policy; however, those obligations fall also to their current electorates, and should inevitably lead them to adaptation and conditioned carbon pricing. In fact, Canada should challenge its fellow G7 countries to $170/tonne by 2030, if not sooner — a simple bright line demand.

Again, no country has more to gain from a reinvention of the UN process than Canada does. No country should press harder than Canada on advancing these concepts of adaptation and pricing, instead of committing haplessly to the current economically self-destructive course. When will Canada rediscover its real interests?

○ ○ ○

Canada's agenda on climate should be premised on the following realities and actions:

- The world will approach a 3°C increase in average global temperature regardless of whether developed economies of the G7 attempt to effect decarbonization. China and India have de facto confirmed this reality.
- The United States, at the federal level, cannot enact climate legislation that would begin their process of mandated decarbonization. There is no political consensus of the kind that can deal with the reality of their legislative process.
- Canada should not attempt to progress decarbonization in the face of these realities, but rather to become an advocate of uniform, properly conditioned carbon pricing across the G7, which, if implemented, will ensure a more efficient low-carbon transition and deal with the issue of free riders such as China and India. Uniform carbon pricing alone can assure Canada that its efforts and sacrifices are not disproportionate relative to the United States and other key trading partners.
- Moreover, if Canada persists with decarbonization domestically, a policy that will have no impact on global hydrocarbon demand or emissions, it will only impose on itself massive economic costs, most of which will be imposed on western Canada and especially Alberta, something that will lead to ever more fracture within the country. This prospect is ever more plausible given the reaction that erupted in early 2022, when the federal government tried to impose vaccine mandates long past any marginal utility had expired. One can only hope such actions will not be repeated in the context of climate policy extremism.

# Afterword

This book was written for a world that was not on verge of genocide in Ukraine. A world where the West did not prove itself incapable of rising immediately to prevent that atrocity. A world where, instead, nations could still rationally assess a global risk and rationally find the optimal means of dealing with it.

It is ironic that so much had been made of the "existential crisis" posed by climate change when, in fact, the entire world has been forced to contemplate the prospect of a genuine existential crisis — the possibility of nuclear annihilation coming about as consequence of this conflict.

The events in Ukraine emphasize ever more starkly the reality of the dependence of modern economies on hydrocarbons and the vulnerability of a Europe that has put itself in the unconscionable position of having to rely on the Russian supply of hydrocarbons. Despite the token bans on the import of Russian oil imposed by the United States and Canada, Europe cannot bring itself to cease the importation of Russian hydrocarbons, even as they are essentially funding the genocide. And at the same time, the West has proven itself incapable of taking sufficient military intervention to actually halt the carnage in Ukraine. An incredible irony. Why is this the case? Concerns about domestic inflation? Concerns about having to remove policy constraints on increasing domestic supplies of hydrocarbons? The risk of provoking the Putin regime to expand the war beyond Ukraine or resort to nuclear weapons? All valid to a point, but the horrific slaughter of the innocent people of Ukraine continues.

The logic behind decarbonization becomes fundamentally undone as Western democracies rediscover that their security interest is equivalent to greater domestic hydrocarbon self-sufficiency.

But of course, delusions die hard. The delusion that decarbonization can "work," that we can cease emitting GHGs and still be able to maintain our existing standard of living. That peaceful coexistence is possible with entities that are so obviously antithetical to any of the values that have defined what one could call Western civilization. Or that sacrificing Ukraine and building more windmills is a tenable course of action.

# Acknowledgements

Naomi Lewis and Dominic Farrell for their skill and guidance. Dundurn Press for their support in publishing *Carbon Change*. And my wife, Maureen, for her wisdom and encouragement.

# Notes

1 Adam Baylin-Stern and Niels Berghout, "Is Carbon Capture Too Expensive?" *IEA* (February 17, 2022), iea.org/commentaries/is-carbon-capture-too-expensive.
2 Robert Bryce, "Maine Voters' Rejection of Transmission Line Shows Again How Land-Use Conflicts Are Halting Renewable Expansion," *Forbes* (November 5, 2021), forbes.com/sites/robertbryce/2021/11/05/maine-voters-rejection-of-transmission-line-shows-again-how--land-use-conflicts-are-halting--renewable-expansion.
3 Ramin Alahdad, Jiya Hai, Guy Holburn, and Brian Rivard, *Energy in Canada: A Statistical Overview* (London, ON: Ivey Energy Policy and Management Centre, 2020), ivey.uwo.ca/media/3792944/iveyenergycentre_policybrief_dec2020_energyinca_overview_editedjan13.pdf.
4 "Trudeau: Governments Grant Permits, Communities Grant Permission," CBC News (March 1, 2016), cbc.ca/player/play/2684686536.
5 "Canada," Climate Action Tracker, climateactiontracker.org/countries/canada.
6 Environment and Natural Resources, "Pan-Canadian Framework on Clean Growth and Climate Change," Government of Canada, canada.ca/en/services/environment/weather/climatechange/pan-canadian-framework.html.
7 "Coronavirus in the U.S.: Latest Map and Case Count," *New York Times*, nytimes.com/interactive/2021/us/covid-cases.html.
8 Jonas Herby, Lars Jonung, and Steve H. Hanke, "A Literature Review and Meta-Analysis of the Effects of Lockdowns on Covid-19 Mortality," *SAE* 200 (January 2022), sites.krieger.jhu.edu/iae/files/2022/01/A-Literature-Review-and-Meta-Analysis-of-the-Effects-of-Lockdowns-on-COVID-19-Mortality.pdf.

# Suggested Reading

Alahdad, Ramin, Jiya Hai, Guy Holburn, and Brian Rivard. *Energy in Canada: A Statistical Overview.* London, ON: Ivey Energy Policy and Management Centre, 2020. ivey.uwo.ca/media/3792944/iveyenergycentre_policybrief_dec2020_energyinca_overview_editedjan13.pdf.

Baylin-Stern, Adam, and Niels Berghout. "Is Carbon Capture Too Expensive?" *IEA* (February 17, 2022). iea.org/commentaries/is-carbon-capture-too-expensive.

Bryce, Robert. "Maine Voters' Rejection of Transmission Line Shows Again How Land-Use Conflicts Are Halting Renewable Expansion." *Forbes* (November 5, 2021). forbes.com/sites/robertbryce/2021/11/05/maine-voters-rejection-of-transmission-line-shows-again-how--land-use-conflicts-are-halting--renewable-expansion/?sh=2362ab5668e8.

———. *A Question of Power: Electricity and the Wealth of Nations.* New York: Public Affairs, 2020.

"Canada." Climate Action Tracker. climateactiontracker.org/countries/canada.

Carney, Mark. *Value(s): Building a Better World for All.* Oxford: Signal, 2021.

Choudhary, Tushar. *Critical Comparison of Low-Carbon Technologies: A Practical Guide to Prioritizing Energy Technologies for Climate Change Mitigation.* Self-published, 2020.

"Coronavirus in the U.S.: Latest Map and Case Count." *New York Times.* nytimes.com/interactive/2021/us/covid-cases.html.

Environment and Natural Resources. "Pan-Canadian Framework on Clean Growth and Climate Change." Government of Canada. canada.ca/en/services/environment/weather/climatechange/pan-canadian-framework.html.

Gates, Bill. *How to Avoid a Climate Disaster: The Solutions We Have and the Breakthroughs We Need.* New York: Knopf, 2021.

Herby, Jonas, Lars Jonung, and Steve H. Hanke. "A Literature Review and Meta-Analysis of the Effects of Lockdowns on Covid-19 Mortality." *SAE* 200 (January 2022). sites.krieger.jhu.edu/iae/files/2022/01/A-Literature-Review-and-Meta-Analysis-of-the-Effects-of-Lockdowns-on-COVID-19-Mortality.pdf.

"In-Depth Q&A: The IPCC's Sixth Assessment on How Climate Change Impacts the World." *Carbon Brief* (February 28, 2022). carbonbrief.org/in-depth-qa-the-ipccs-sixth-assessment-on-how-climate-change-impacts-the-world.

"In-Depth Q&A: The IPCC's Sixth Assessment Report on Climate Science." *Carbon Brief* (August 9, 2021). carbonbrief.org/in-depth-qa-the-ipccs-sixth-assessment-report-on-climate-science.

IPCC. *Climate Change 2021: The Physical Science Basis.* Cambridge: Cambridge University Press, 2021. ipcc.ch/report/sixth-assessment-report-working-group-i.

———. *Climate Change 2022: Impacts, Adaptation and Vulnerability* Cambridge: Cambridge University Press, 2022. ipcc.ch/report/sixth-assessment-report-working-group-ii.

Kahn, Matthew. *Adapting to Climate Change: Markets and the Management of an Uncertain Future.* New Haven: Yale University Press, 2021.

Koonin, Steve. *Unsettled: What Climate Science Tells Us, What It Doesn't, and Why It Matters.* Dallas: BenBella Books, 2021.

Lomborg, Bjørn. *False Alarm: How Climate Change Panic Costs Us Trillions, Hurts the Poor, and Fails to Fix the Planet.* New York: Basic Books, 2021.

———. *The Skeptical Environmentalist: Measuring the Real State of the World.* Cambridge: Cambridge University Press, 2001.

McConaghy, Dennis. *Breakdown: The Pipeline Debate and the Threat to Canada's Future.* Toronto: Dundurn, 2019.

———. *Dysfunction: Canada After Keystone XL.* Toronto: Dundurn, 2017.

Nordhaus, William. *The Climate Casino: Risk, Uncertainty, and Economics for a Warming World.* New Haven: Yale University Press, 2015.

———. *A Question of Balance: Weighing the Options on Global Warming Policies.* New Haven: Yale University Press, 2014.

Pitron, Guillaume. *The Rare Metals War: The Dark Side of Clean Energy and Digital Technologies.* N.p.: Scribe U.S., 2020.

Schellenberger, Michael. *Apocalypse Never: Why Environmental Alarmism Hurts Us All.* New York: Harper, 2020.

"Trudeau: Governments Grant Permits, Communities Grant Permission." CBC News (March 1, 2016). cbc.ca/player/play/2684686536.

Yergin, Daniel. *The New Map: Energy, Climate, and the Clash of Nations*. New York: Penguin, 2020.

# Index

# About the Author

Dennis McConaghy, a native Albertan, enjoyed an extensive career as one of Canada's pre-eminent energy executives over the last thirty years, being directly involved in the development of some of the country's most iconic energy projects, including Keystone XL, Coastal GasLink, Mackenzie Valley Gas Pipeline, and Energy East.

Since retirement, he has been a noted author and commentator about Canadian energy and climate policy. He is the author of *Dysfunction: Canada After Keystone XL* and *Breakdown: The Pipeline Debate and the Threat to Canada's Future*, winner of the 2019–2020 Donner Prize.

Dennis has consistently advocated for a better balance in how Canada fully captures the economic value of its hydrocarbon endowment while achieving a climate policy that is both credible and commensurate to the reality of its position in the global economy.